Bioinformation Worlds and Fu

This book sets out to define and consolidate the field of bioinformation studies in its transnational and global dimensions, drawing on debates in science and technology studies, anthropology and sociology. It provides situated analyses of bioinformation journeys across domains and spheres of interpretation. As unprecedented amounts of data relating to biological processes and lives are collected, aggregated, traded and exchanged, infrastructural systems and machine learners produce real consequences as they turn indeterminate data into actionable decisions for states, companies, scientific researchers and consumers. Bioinformation accrues multiple values as it transverses multiple registers and domains, and as it is transformed from bodies to becoming a subject of analysis tied to particular social relations, promises, desires and futures. The volume harnesses the anthropological sensibility for situated, fine-grained, ethnographically grounded analysis to develop an interdisciplinary dialogue on the conceptual, political, social and ethical dimensions posed by bioinformation.

EJ Gonzalez-Polledo teaches anthropology at Goldsmiths, University of London. They are the author of *Transitioning: Matter, Gender, Thought*, and are currently developing research on global open biology movements and global histories of bioinformation.

Silvia Posocco is an anthropologist based at Birkbeck, University of London. Posocco is the author of *Secrecy and Insurgency: Socialities and Knowledge Practices in Guatemala*. Current projects include research on the archives of transnational adoption in the aftermath of genocide as well as new collaborative work on global histories of bioinformation.

Routledge Studies in Anthropology

https://www.routledge.com/Routledge-Studies-in-Anthropology/book-series/SE0724

Bioinformation Worlds and Futures

Edited by
EJ Gonzalez-Polledo and Silvia Posocco

Routledge
Taylor & Francis Group

LONDON AND NEW YORK

First published 2022
by Routledge
2 Park Square, Milton Park, Abingdon, Oxon OX14 4RN

and by Routledge
605 Third Avenue, New York, NY 10158

Routledge is an imprint of the Taylor & Francis Group, an informa business

© 2022 selection and editorial matter, EJ Gonzalez-Polledo and Silvia Posocco; individual chapters, the contributors

British Library Cataloguing-in-Publication Data
A catalogue record for this book is available from the British Library

Library of Congress Cataloging-in-Publication Data
A catalog record has been requested for this book

ISBN: 978-0-367-40945-6 (hbk)
ISBN: 978-1-032-14082-7 (pbk)
ISBN: 978-0-367-81003-0 (ebk)

DOI: 10.4324/9780367810030

Typeset in Sabon
by Taylor & Francis Books

Contents

Figures

Acknowledgements

The editors acknowledge the generous support of the following institutions for our collaborative and individual bioinformation-related projects: Gonzalez-Polledo and Posocco, 'Data Worlds and Futures: Archives, Bioinformation and Evidence', Wellcome Trust Small Grant, 2019; Gonzalez-Polledo and Posocco, Birkbeck/Wellcome Trust Institutional Strategic Support Fund (ISSF), Conference and Symposia Fund, 2019; Posocco, 'Person, Substance, Bodily Capacity: Transnational Adoption and Surrogacy in Times of Violence and Crisis', Birkbeck/Wellcome Trust Institutional Strategic Support Fund (ISSF), 2016–2017. We are grateful to Katherine Ong, Routledge Anthropology, for the enthusiasm she expressed for the idea of this book project and the authors for accepting our invitation and persevering with the project in Covid-19 times.

Contributors

Anisha Chadha is an anthropologist whose work explores the production of biomedical technologies in South Asia and the US, particularly focusing on how caste and gender dynamics inform sites of clinical experimentation. Her research and teaching interests include critical studies of biomedicine, and the intersection of ethnographic methods with contemporary design and engineering training. She is currently a PhD candidate at New York University.

Resto Cruz is Lecturer in Social Anthropology at the University of Edinburgh. His work centres on how lives and relationships unfold over time, the traces that accumulate in their wake, and how they are shaped by, and generate, wider historical transformations. His interests include social and geographical mobility, siblingship, birth cohort studies, life course epidemiology, the Philippines, and the UK.

Laura Fenton is Research Associate in the School of Health and Related Research at the University of Sheffield. She completed a PhD in sociology at the University of Manchester in 2018 on the place of alcohol in the day-to-day lives of three generations of British women born between the 1940s and the 1990s. Her research interests include youth, gender, the life course, and creative biographical methods.

EJ Gonzalez-Polledo teaches anthropology at Goldsmiths, University of London. They are the author of *Transitioning: Matter, Gender, Thought*, and are currently developing research on global open biology movements and global histories of bioinformation.

Anna Jabloner teaches in the Department of Anthropology at Harvard University. She is an anthropologist with research interests in genetics, biomedicine, identity, ethnographies of the United States, and feminist, queer, and critical race theory. Her work explores different uses of, and imaginations about, genetic technologies across clinic, industry, law and criminology in the US.

Jongmi Kim is Senior Lecturer at Coventry University, UK. Jongmi's research covers a wide range of subjects from postcolonial media studies to constructing women's identities through cosmetic surgery in South Korea. Recently, she has extended her research interests to the area of co-shaping mobile technology and media representation in Japanese society.

Kiheung Kim is Associate Professor at Pohang University of Science and Technology (Postech), South Korea. Kiheung has conducted a series of research projects on the sociological relationship between outbreaks of infectious diseases like BSE, MERS and Foot-and-Mouth disease and social transformations.

Peter C. Little is Associate Professor of Anthropology at Rhode Island College. He is the author of *Toxic Town: IBM, Pollution, and Industrial Risks* and his forthcoming book is *Burning Matters: Life, Labor, and E-Waste Pyropolitics in Ghana*. He is currently working on a book entitled *Platforms, Pandemics, and Plunder: New Intersections of Technopower and Global Political Ecology*.

Sylvia McKelvie is an interdisciplinary researcher and writer. She holds degrees in Sociology from the University of British Columbia and the London School of Economics and Political Science. Her main research interests are urban politics, public health, STS, critical race theory and feminist praxis.

Mark Maguire is Dean of Maynooth University Faculty of Social Sciences. He researches counterterrorism and airport policing. Recent publications include *Getting through Security: Counterterrorism, Bureaucracy and a Sense of the Modern* with David A. Westbrook.

Eileen Murphy is a doctoral fellow in Copenhagen Business School studying the politics and service dimensions of European large-scale data infrastructure. She has contributed to projects on automated border control and EU security as a researcher in Trinity College Dublin and the Centre for Irish and European Security (CIES).

Tahani Nadim is a junior professor at the Humboldt University of Berlin in a joint appointment with the Museum für Naturkunde Berlin. She heads the department Humanities of Nature, which is dedicated to an interdisciplinary investigation of the politics of nature. Her research focuses on the datafication of nature and questions of political accountability in digitizing collection.

Silvia Posocco is an anthropologist based at Birkbeck, University of London. Posocco is the author of *Secrecy and Insurgency: Socialities and Knowledge Practices in Guatemala*. Current projects include research on the archives of transnational adoption in the aftermath of genocide as well as new collaborative work on global histories of bioinformation.

Adrian Van Allen is an Associate Researcher in Anthropology at the Smithsonian National Museum of Natural History, and the California Academy of Sciences. Currently an Abe Fellow of the Social Science Research Council, she is a cultural anthropologist who studies museums as technologies for organizing relationships between people, places, materials, and interests.

Noah Tamarkin is Assistant Professor of Anthropology and Science & Technology Studies at Cornell University and a research associate at the Wits Institute for Social and Economic Research (WISER) at the University of the Witwatersrand in Johannesburg, South Africa. His research examines how DNA transforms power and politics as it becomes unevenly part of everyday life through technologies like ancestry testing and criminal forensics. He is the author of *Genetic Afterlives: Black Jewish Indigeneity in South Africa*. His work has also appeared in *Cultural Anthropology, American Anthropologist, Catalyst: Feminism, Theory, Technoscience, History and Anthropology*, and *Annals of the American Academy of Political and Social Science*. His current ethnographic research, supported by the US National Science Foundation, the Wenner Gren Foundation, and the Ohio State University's Criminal Justice Research Center, examines the introduction and implementation of legislation to expand South Africa's national forensic DNA database.

Penny Tinkler is Professor of Sociology and History at the University of Manchester. She has published extensively on the 20th-century history of girls/women, and on photographic methods. She currently leads the ESRC-funded project 'Transitions and Mobilities: Girls Growing Up in Britain 1954–76 and the Implications for Later-Life Experience and Identity'.

1 Bioinformation worlds and futures

An introduction

EJ Gonzalez-Polledo and Silvia Posocco

Bioinformation refers to information that is derived from biological organisms or that describes biological processes and lives (Parry and Greenhough 2018). Its emergence as a concept across academic disciplines and domains of practice marks transitions from body to data, biological substance to information, and archives to datasets. Linking life and information in the digital age, bioinformation has become newly entwined in social relations. As unprecedented amounts of data relating to biological processes and lives are collected, aggregated, traded and exchanged, infrastructural systems and machine learners produce real consequences as they turn indeterminate data into actionable decisions for states, companies, scientific researchers and consumers. The uptake of ancestry DNA and direct-to-consumer genetic testing, for example, has risen exponentially in the past five years as individuals seek to explicate their own genealogies of historical and contemporary relatedness. This rise can be explained by three factors that centre multiple values and changing understandings of bioinformation. Firstly, populations undergoing direct testing have increased across the world; second, the aggregation of genetic data produced by the corporate consolidation of DNA testing provision and the increased interoperability of formerly discreet platforms are creating new, and much more tightly intertwined landscapes of bioinformational exchange. In 2019, the DNA of over 26 million people was accessioned to the world's leading testing companies, whose user numbers in 2019 outstripped all previous years combined (Regalado 2019). There is also much more traffic of information between providers as individuals transfer raw DNA samples and derivative bioinformation is uploaded to one platform, such as Ancestry.com, to others in search of answers to health, paternity or other questions.

As big data and algorithmic processes learn to predict every aspect of social life with ever more granularity, anthropology must turn to framing the entangled histories and temporalities which bioinformation connects and actualises. The global Covid-19 pandemic has made visible how bioinformation is transforming biological citizenship (Rose and Novas 2004, Petryna 2004) at a time when governments and public health bodies seek to ease lockdown restrictions and return to safe levels of social interaction. As genetic and antibody information becomes essential to understand the pandemic origins and track transmission and immunity in human populations (Forster et al. 2020, Ferretti et al. 2020),

DOI: 10.4324/9780367810030-1

bioinformation becomes an essential interface that reconfigures relations between citizens, governments, and institutions. As testing and other forms of intervention evince (Marres and Stark 2020), bioinformation is constitutive of – and not external to – forms of governance and sovereignty, but its analysis in public health contexts can have detrimental consequences for differently gendered, sexualised and racialised subjects and populations, who are further differentiated on the grounds of genetic, illness, HIV- and citizenship status. Bioinformation analysis, as practice and figuration, is therefore also enmeshed in routinised forms of structural violence, modes of precarious existence, illness and premature death that are linked to deprivation and gendered and racialised inequalities.

In this context, new questions are emerging about the shaping and social impact of bioinformation extraction, storage and circulation. The aggregation of multiple sources of genetic data similarly generates significant challenges in forensics, producing unprecedented tensions between national and supranational criminal justice systems and enforcement cultures as they vie for control of identifying information. For example, the aggregation of DNA in global genetic databases that can be cross-matched with private company data significantly increases the probability of discovering potentially life-changing information about kinship and ethnicity. Genomic mapping companies promise to transform outcomes following a life-changing diagnosis, but unregulated availability of analytic services has also led to data breaches, misguided decisions, and unwanted harm. In the current context, and as contributors to this volume show, the readability, transparency and legibility of bioinformation are collectively transformed by historically unprecedented levels of interoperability between public and private actors across the Global North and South. In the age of global infrastructures, bioinformation is routinely donated, exchanged, traded or extracted under diverse arrangements and transactional forms that involve persons, institutions and biotechnologies. Indeed, the global reach of bioinformation infrastructures make bioinformation a 'global social fact' – a form of relationality that links multiple localities, infrastructures, scientific cultures and interests. While access to personal bioinformation analysis is facilitated by the availability of genetic sequencing, the unlicensed exchange of genetic bioinformation between commercial, research and clinical environments emerges as a fundamental challenge as donors remain largely unaware of the consequences of global bioinformation traffic, marketisation and aggregation. As differences in the type, quality and integration of existing services widens, bioinformation demands new conceptual approaches, intensive transnational empirical investigation, and the institution of global conversations between diverse stakeholders.

Anthropological studies of biomedicine and bioethics have offered incisive accounts of how some of these transitions map onto the attribution or disavowal of personhood in medical contexts or how different epistemic communities might figure relations and belonging through these new technologies (Tamarkin 2014). They show how cultures of expertise bridge expert and lay divides, as the 'new genetics' (Pálsson 2007, 2008) offer new visual and mapping imaginaries for figuring relatedness, property and value. New 'biosocialities' (Rose 2007) also emerge, as bioinformation in its genetic iteration becomes an area of concern for publics as diverse as Indigenous communities and geneticists (TallBear 2013), patient groups (Shaw 2009), and sheep (Franklin 2007) and horse breeders (Cassidy 2002). Through new conceptual and ethnographic analyses of the purchase, ethics and politics of bioinformation, this volume unpacks questions of access and interpretation, taxonomic and individual identification, appropriation and reproduction. With Gregory Bateson, we propose to understand bioinformation as a 'pattern that connects'. Through bioinformation, individuals become enmeshed in unexpected relations as they acquire knowledge that rewrites previous understandings of their identity, kinship ties, ancestry, health status, and even prospective criminality. Taking bioinformation as an object of ethnographic analysis also makes perceptible the multiple values that bioinformation accrues as it travels from bodies to data in particular databases, archives and infrastructures. Moving from analysing systems, infrastructures and ecologies to understanding the cultures, practices and contexts that shape their development, this volume aims to answer a central question: How do cultural elements shape public understandings of life and death through bioinformation? In this context, we look at how well bioinformation travels across fields and domains of practice, drawing the significance of their social lives and afterlives for the public domain of science. We are particularly concerned with science as a practice of truth-making in the pursuit of wellbeing and justice, and the establishment, maintenance and transmission of 'cultures of evidence' across multiple domains of expertise. Drawing on ethnographic research in forensic archives, museums, biorepositories, medical technologies prototype manufacturing, e-waste management, airport biosecurity surveillance and forensic research centres, we focus on shifts in data materialities and data practices that enable new connections and disconnections between epistemic communities and publics. These relations, we argue, are constitutive of bioinformation throughout the life cycle, as biological specimen and resources are progressively disembodied and re-contextualised across multiple infrastructures from before birth to after death.

Life, death, data

Life and death are at the heart of contemporary debates in social theory. The proximity and interconnections between processes of living and dying, the uneven and unequal distributions of life chances, and the proliferation of precarious forms of existence in the context of social exclusion, dispossession, violence and crisis have generated considerable debate and critical reflection. Foucault's theorisations of biopower and biopolitics in particular have offered a theoretical register to grapple with relations between institutions, forms of knowledge and expertise, and practices of 'letting live' and 'making die' (Foucault 1981, 2001, 2003). This field has progressively highlighted the ways in which forms of vulnerability, exposure and expendability are constitutive of – and not external to – forms of governance and sovereignty, with deadly consequences for differently gendered, sexualised, racialised, and HIV ser-ostatus marked subjects and populations (Agamben 1998, Comaroff 2007, Povinelli 2011). Whilst Foucault focused on biopower as 'a power to foster life or disallow it to the point of death' (Foucault 1981: 138), a sustained emphasis on sovereign power as fundamentally concerned with death-making has emerged. This scholarship has highlighted progressive, routinised forms of structural violence and how 'letting die' connects to modes of precarious existence and the normalisation of extreme suffering. In this context, illness and premature death are linked to deprivation and gendered and racialised forms of marginalisation (Biehl 2001, Farmer 1996). In the midst of a data revolution transforming social worlds, data infrastructures and analytics are fast transforming what it means to live and die. Infrastructures have brought forward new ways of doing and believing in evidence, as simulation techniques make them-selves 'easy to love and difficult to doubt' (Turkle 2009: 7, see also Kennedy 2016) by scientists and general publics.

Predictive analytics and machine learning based on a ubiquitous process of datafication have foreseen 'the end of theory' (Anderson 2008), replacing 'tra-ditional' classification practices and methods in the sciences with process sensi-tive, 'real-time' data-driven analytics (Mayer-Schönberger and Cukier 2013). In this context, data infrastructures – tools for storing, processing and analysing data – are becoming ever more central to the functioning of social worlds. Data infrastructures not only include data registers, but technological processes and organisational principles that detail how data can be accessed and processed, as well as the principles that govern human-technology interactions. In this con-text, changing notions of public and private, individual and collective, infor-mation and evidence accelerated by data-driven science are of significant consequence for the lives and deaths of individuals (Nisa 2016). However, the relation between the promise of these technologies and how they can make a difference in research, policy and service delivery is still far from settled.

This volume argues that living and dying are underpinned by geographies and infrastructures of connectivity, prosperity and wellbeing, and of risk, toxicity and exposure, which are generative of new tensions and frictions in the body politic (Harvey and Knox 2015, Harvey, Jensen and Morita 2017, Mitman, Murphy and Sellers 2004). Infrastructure studies place emphasis on the 'socio-material' status of infrastructures, and the consequent instability, dynamism and differential impact of the infrastructures and computational architectures that subtend life and death. Building infrastructures and design of analytic interactions can further economic inequalities through practices of targeting, prediction and ranking. Infrastructures extend life- and death-making practices into emergent domains of social practice which, whilst offering opportunities for connectivity, sociality and identification, also entail differential burdens of risk and vulnerability. The chapters collected in this volume analyse a range of biorepositories, archives, museums, forensic laboratories and biobanks to con-sider how, in the digital age, bioinformation raises urgent questions about how computational designs, architectures and digital cultures may be reorganising life and how governments, regulators and constituencies may respond to the increasingly central role of bioinformational technologies in the everyday.

As biosocialities (Rabinow 1996), the anthropology of the 'new genetics' (Pálsson 2007, 2008), and the cultures and publics of scientific infrastructures rearticulate the relation between the biological and the social in theory and social practice, biosocialities exemplify how social understandings of genetic information may give rise to novel configurations of subjectivity and belonging, generating new social identities as individuals grapple with the implications and possibilities inherent in biological data and genetic analytics. Bioinformation – derived from the analysis of physical or biological characteristics of a person – is now redefining approaches to security, surveillance and governmentality informing the development of new logics and techniques of managing economic, social and political relations. For example, health researchers have stressed how taxonomies of health and illness and related diagnostic categories underpin the emergence of genetic identities, advocacy and activism (Gibbon and Novas 2008, Taussig 2009). They have also charted the inequalities and exclusions that map onto genetic and health determinants. The anthropology of 'the new genetics' has been increasingly concerned with highlighting the pitfalls of objectifying and commodifying users, foregrounding instead the biosocial relations that sustain genomic services and the establishment and maintenance of biobanks, in the context of the emergence of personalized medicine and the economic exploitation of bioinformation (Ventura Santos and Wade 2014, TallBear 2013, Prainsack et al. 2014). More fundamentally, the new genetics entails grappling with the inherent partiality of genetic knowledge – e.g. DNA information – and a sustained engagement with the articulations of social meanings which are elicited by the incompleteness of scientific information (Franklin 2003).

We suggest moving beyond the framework of socialities of diagnosis and sociologies of expectation, and propose that analyses of bioinformation worlds ought to encompass multi-scale investigations of the infrastructures, cultures of evidence, and scientific practices that emerge as corporeal information is turned into data. The volume takes forensic archives, research and innovation centres, biorepositories, museums and biobanks as ethnographic field sites to examine the making of life and death through data, and as key locales where bioinformation can be tied to data publics – constituencies seeking access and participation in the making and interpretation of bio- and genetic information. Through a focus on how data is handled, processed and interpreted by different communities – e.g. scientists, archivists, technical staff, software experts, anthropologists, and user groups – the essays collected in this volume draw on these established perspectives to ask new questions that expand the debate in novel directions, looking to transform the understanding of the practices, value and values underlying decisions by data.

Patterns and relations

Bioinformation derived from the analysis of physical or biological characteristics of a person is now redefining approaches to security, surveillance and governmentality, informing the development of new logics and techniques of managing economic, social and political relations. For example, health researchers have stressed how taxonomies of health and illness and related diagnostic categories underpin the emergence of genetic identities, advocacy and activism. Bioinformation research takes the forensic archive, the forensic research centre, the biorepository and the biobank as ethnographic field sites to examine the making of life and death through data, and as key locales where bioinformation can be tied to data publics – constituencies seeking access and participation in the making and interpretation of bio- and genetic information. Debates on genomic archives and post-archival genomics shed light on the shifting constitutions of bioinformation that move from biology to logistics, a shift enabled by the availability and accessibility of genomic sequencing technologies. Through modelling, visualisations and infrastructure maps, computational architectures are shown to be tied to how these complex tensions between presence and absence may persist in digital archives. An anthropology of bioinformation must encompass data structures and architectures, processes of set up and operation of local networks, infrastructures and cloud technologies, as well as the operations of algorithms used in data processing. This approach builds on philosophies of computation based on combining qualitative and quantitative approaches, to answer targeted (explanation oriented) and meta questions, to map how infrastructures operate and think, and how they help us make sense of data, understand and explain them. Cloud geographies (Amoore 2018) currently pose significant stakes for governments, corporate health providers and health publics, as their management, use and access has become a critical asset of health systems. Ethnographic methods can probe whether and how infrastructures are invested with powers to scale interventions and reduce uncertainty, for example, or consolidate taxonomic orders through which species boundaries are devised, regulated and governed – as Chapter 2, by Tahani Nadim, shows.

Anthropological accounts of bioinformation mobilise formative debates in anthropological theory which have foregrounded how categories and relations emerge at the intersections of knowledge and social practice. The interplay between scientific or expert knowledge and situated experience is central to the study of bioinformation, as individuals and communities actively engage in the making and remaking of 'the politics of life itself' (Rabinow and Rose 2006, Rose 2001). In turn, domains of expert knowledge have been shown to be rooted in malleable milieus, where lay discourse often shapes research questions and practices. The history of the gendering of the X and Y chromosomes (Richardson 2013) can serve as a notable example of this dynamic. Medical anthropology has extensively documented these processes, notably with reference to people's understandings and interactions with specific medical technologies, diagnostic categories, and expert theories and models (Lock 2017, Thompson 2005). Ethnographic approaches to the study of biosocialities are, again, instructive here. Re-articulating the relation between the biological and the social in theory and social practice, biosocialities exemplify how social understandings of genetic information, for example, may give rise to novel configurations of subjectivity and belonging, generating new social identities as individuals grapple with the implications and possibilities inherent in derivative analyses of biological data.

Anthropological perspectives extend this discussion beyond an account of the connection and interplay between expert and lay domains, knowledge and social practice, science and culture. They re-centre notions of 'patterns' alongside 'relations' to address the urgent questions posed by the contemporary (re)emergence of bioinformation as a research object and site of social and cultural anxieties. A rapprochement with anthropological theory concerned with systems and patterns, from the work of Gregory Bateson to Lévi-Strauss, through to Strathern's (2020) latest re-conceptualisation of the notion of 'relation', may be an apt starting point for anthropologies of bioinformation. In *Mind and Nature* (1979), Bateson offers an expansive discussion of the relationship between epistemology, or *knowing*, in Bateson's own terms, and the world. Bateson invokes a 'wider knowing which is the glue holding together the starfishes and sea anemones and redwood forests and human committees' (1979: 5), suggesting that a broad set of terms might be needed to imagine and understand what enables the coalescence of human and non-human figures. Bateson recasts longstanding debates about the status of the distinction between the categories of the 'human' and the 'non-human', '*creatura* (the living) and *pleroma* (the non-living)' (1979: 7), and the contours and boundaries of 'life' and proposes 'the pattern that connects' as a way to think through some of these questions and problems:

> *The pattern which connects* … What pattern connects the crab to the lobster and the orchid to the primrose and all the four of them to me? And me to you?
>
> (1979: 7)

The pattern which connects is, for Bateson, dynamic and always already in context, that is, 'a pattern through time' (Bateson 1979: 14) and across scales of relation and abstraction through which 'the living world' can be figured. It is our contention that bringing Bateson's pattern which connects to bear on the analysis of bioinformation is particularly apt, given Bateson's own interest in the convergence of life and information. Bateson's arguments in *Mind and Nature* prefigure many of the questions that emerge in contemporary research on bioinformation, as patterns of interconnectedness come into renewed focus and raise urgent questions about how exactly we are to think about persons, relatedness and collectivities through bioinformational relations in which the entanglement of chemical and informational processes defines the process of living.

Indeed, the idea that biological elements such as the gene contained information found resonance with cybernetic theories of communication. Von Newmann's definition of a self-producing machine appeared in his path-breaking paper 'General and Logical Theory of Automata', presented in Pasadena, California, on September 20, 1948, in which he famously equated DNA and software. Von Neumann, an outsider in the field of genetics, ventured in this paper to offer a mathematical theory of artificial autonomy applied to 'natural organisms' – a task nested under a broader aim to compare complex living organisms with artificially generated automata, a problem that von Neumann related to units and fragments which functioned individually within a broader understanding of how fragments are organised into wholes, the function of which is revealed and expressed in the individual elements. Von Neumann hypothesised that axiomatising the behaviour of their elements might lead to understanding the functional characteristics of systems, defined through how unambiguously they responded to stimuli by deploying analogic thinking in linking 'natural' and artificial life. To von Neumann, artificial automata had similarities with the central nervous system, and, as such, they were needed when the complexity of a task involved functionality beyond elementary operations.[1] Von Neumann understood living organisms to have elements of both digital and analogue systems, defined for demonstration purposes as digital automata: not only were they composed by elementary units whose functions mirrored those of automated elements, but the system responded to stimuli which directed the flow of energy from the element in question to the source (McMullin 2000). With this description, von Neumann opened up a new grammar to understand the molecular differences upon which organisms are built, interpreting biology through digital parameters based on yes-no responses to information.

For Bateson, the pattern that connects enabled the establishment of links between, for example, chemical pathways and structures, parts and wholes, ethological and biological process (see Bateson 1979). Likewise, an important dimension of bioinformation as a conceptual device is the totalising vista it produces and enables. The pattern that connects is an idiom that speaks to claims often attached to bioinformation as enabling what Haraway (1988) termed the 'god trick' of a view from nowhere, enabling totalising vistas which often fit governmental rationales and exigencies. Bioinformation holds a promise of total connectivity and totalising knowledge which Bateson aptly expressed with reference to the pattern that connects, as follows:

> How do ideas, information, steps of logical or pragmatic consistency, and the like fit together? How is logic, the classical procedure for making chains of ideas, related to an outside world of things and creatures, parts and wholes? … What has to be investigated and described is a vast network

matrix of interlocking message material and abstract tautologies, premises, and exemplifications.

(Bateson 1979: 19–20)

Patterns that connect often stand for infinity, their heuristic qualities appearing as objects, figures or interactions in the social world that are returned to us, recursively, and 'seen twice', or an infinite number of times (Riles 1998). An anthropological approach may ask 'whose patterns are these?' (Riles 1998: 394). This question draws attention to the importance of understanding context and location, but also extends the discussion to a consideration of how information may become property and transaction (Hirsh and Strathern 2006). Bioinformation is assembled through social, political and technical material-semiotic entanglements. The historical emergence of bioinformation worlds as objects of analysis can be traced in reference to a number of scientific developments, such as the rise and marketisation of genomic science, but also practices of making sense tied to state apparatuses (Koopman 2019). Indeed, Susan Oyama (2000) laid out the significance of information beyond genetic prefiguration, placing information beyond hylomorphic distinctions that sustained definitions of life and death. Information, as a becoming phenomenon, is instead 'dependent on its actual functioning' (Oyama 2000: 3). For example, the notion of 'emergence' has taken new ground in explanations of organisms as collectives, describing how processes and properties result from informational interactions between parts, between parts and wholes, and between systems and environments.

We are specifically interested in global perspectives on technoscientific practices, and particularly the global utopian visions that underpin large scale technological projects. As bioinformation storage and analysis inform policy shifts in this changing field, technologies of bioinformation extraction and interpretation are newly linked to individual and collective bodies by enacting long cultural histories. For example, M'charek (2020) shows that bioinformation not only identifies individuals, but rather develops as 'tentacular' mechanisms of racialisation employed first and foremost in the making and management of populations. Medina's (2011) account of the rise and fall of Project Cybersyn, the nation-wide technological system which would offer a real-time picture of economic production to inform economic policy and management in Allende's Chile, is also instructive in this respect.

While the turn to bioinformation led communication theorists to reflect the textual fabrics of life (Kay 2000, Oyama 2000), for anthropologists and other social scientists it also revealed the social fabrics that underpinned the rise of bioinformation as a key commodity at the core of definitions of life and its futures. Bronwyn Parry's (2004) global ethnography illustrates how the global circulation of bioinformation relies on complex ethical, political and economic entanglements. Bioinformation has become a commodity in a rapidly evolving technoscientific landscape, key to changing relations between humans and technics. Following bioinformation as it turned into a highly prized commodity in an informational economy, Parry tracked the development of new markets and forms of exchange (see also Hoeyer 2013), and demonstrated how the slippery nature of shifting bioinformation and their value forms for different stakeholders made any effective market regulation challenging. At the same time, the biotech industry grew exponentially seeking lower production costs and labour in developing countries, supporting a shift in the biosciences towards data-centric approaches (Leonelli 2016) that anchored science to the development of ever more complex data architectures and infrastructures.

Bioinformation worlds and their futures

In Chapter 2, 'All the data creatures', Tahani Nadim focuses on an analysis of DNA sequences as empirical sites where the concept of species is dynamically articulated. The DNA sequence of a West African dwarf crocodile provides a concrete and yet abstract example of the analytical work DNA sequencing entails, notably in the form of the genetic barcode – a standardised short sequence of DNA. Genetic barcodes are shown to be devices geared towards species identification orchestrated by contemporary fantasies of the total (genetic) archive of natural history. These fantasies have a long history in the knowledge formations and practices of pillaging and capture of Empire. The chapter considerably advances research on bioinformation, by identifying processes and artefacts – and their infrastructural and rhetorical organisation – notably, 'bioinformation pipelines' and 'data moments', two important analytical contributions to the study of bioinformation and 'data-centric biology' (Leonelli 2016). Three 'data moments', or case studies, illustrate how the project of circumscribing 'species' is in fact continuously undermined by the capacity of database records to multiply relations in and through 'an elsewhere' that always displace 'the original referent'. Nadim examines the book of accession of specimens in a department of the Museum für Naturkunde in Berlin, the digitalisation of records and use of Excel sheets and how marginalia are excised in the digital conversion – a point that resonates with the analysis of cohort studies records offered by Cruz, Tinkler and Fenton in Chapter 4. The barcode is tied to claims about the efficiency, accuracy and speed in species identification which illustrate the data practices at play in sieving, layering, smoothing and producing more data through manual and automated operations. Data moments therefore show how barcodes generate, rather than represent, species distinctions.

In Chapter 3, 'Capturing genomes: the friction and flow of bioinformation at the Smithsonian', Adrian Van Allen offers a compelling ethnographic account of how samples are added to the genomic collection in the Smithsonian Institution Biorepository. Van Allen explores the processes of remaking, re-inscribing and removing boundaries between nature and culture in the everyday handling of biomaterials, as animals become specimens in a collection geared towards biodiversity conservation against the horizon of ecological crisis. The tissue samples and the bioinformational objects that derive from them hold the promise of reproducing the future, a potentiality that is not realised or brought to term in the biorepository, but which shapes the social and data practices which unfold in and through the bio-materials. Genomic collecting relies on material practices which create spe-cimens that are made to stand for animals such as the crab and the fish whose tissue samples Van Allen follows through the genomic museum. The creation of the genomic archive illustrates ethnographically what Fortun has called 'care of the data' (Fortun 2019), that is, the range of sensory and scientific procedures that are often overlooked but that are key to the fash-ioning of biodiversity conservation in the museum repository. It is Van Allen's contention that these are also practices through which power oper-ates and varied forms of capture (Parreñas 2018) are actualised. Friction and flow characterise genomic collecting, as bioinformation is captured, frozen and preserved for uncertain futures.

Chapter 4 focuses on an analysis of how bioinformation is entangled in the figuring of kinship in the context of large cohort studies, which underpinned the development of totalising archive projects built around visions that centred person data in governance and policy. Cruz, Tinkler and Fenton follow the 1946 cohort study to ascertain how different types of data gradually resulted in compiling population databases. A cohort birth study presents an opportunity to frame the importance of relations as these emerge and evolve in the way archives materials are assembled, and in the logics, voices and gaps preserved in the archive over time. Cruz, Tinkler and Fenton's chapter is a prime case study of state-led and funded bioinformation processing under research programmes designed to inform public health interventions and health policy at national level. The chapter thus makes an incisive interven-tion in debates that centre bioinformation in the present, by showing that the collection and analysis of bioinformation are part of longer histories of epi-demiological and biomedical research that entangled bodies, information and governance (Koopman 2019, Foucault 1991). 'Salvaging' the archives through creative practices of recomposition and creative interpretation, the chapter fol-lows bioinformation as it is interpreted first hand by the participants in the study, drawing out the significance of the bioinformation in figuring kinship relations and the significance of kinship in the generation and function of bioinformation, cen-tring how kinship, and more specifically parent-child relations, were essential to the constitution and storage of biodata, and how, conversely, databases were modelled around kinship transmissions. These are, then, also histories of institutions, as research bodies are established and then fold, when new prio-rities and technologies emerge to signal the future. The chapter shows that the bioinformation modelling within cohort studies is also underpinned by a need to make sense of biological data as a blueprint for social practice.

Writing about a new role of bioinformation in personalised medical services, in Chapter 5, 'Bioinformation in formation: inventing medical devices in contemporary India', Anisha Chadha offers a rich ethnographic exploration of the place of bioinformation in the making of medtech devices in Delhi. Chadha centres the processes that lead to the fashioning of these devices to capture the nuances, expectations and promises that link bioinformation to consumer medtech devices that make use of technical capacities of mass bioinformation extraction. Chadha's account of bioinformation processing and circulation takes the vantage point of local entrepreneurs. Through ethnographic observation, Chadha opens the relationship between bioinformation, biomedical technologies and the professional identities and innovation cultures populating a dynamic biotech sector in India. Chadha focuses on prototyping as a practice that stages real and imagined relations between human bodies, governmentality, promissory regimes and futures, as prototypes embody the rise of entrepreneurial citizenship with hopes to develop global development opportunities (Lindtner 2020, Irani 2019). As Indian medtech engineers employ prototyping models to conceptualise social dimensions of biomedicine, including the relation between humans and digital devices, bioinformation transfers and circulation defines the process of engineering problem space as a set of evolving relations between biological, heuristic and semiotic materials. These translations are not only consequential in how innovation cultures shape biomedical research and service industries, but pose crucial ethical challenges that frame the value and values of bioinformation.

The next set of chapters address state-led bioinformation processing in criminal justice and surveillance systems. Both McKelvie and Jabloner's contributions show that the field of forensics is often located at the intersections of criminal justice and medicine. McKelvie stresses the ways medical technologies deployed in the aftermath of sexual assaults in the UK often entail intrusive procedures of bioinformation recording. Jabloner, in turn, examines the other end of the forensic pipeline, namely datasets which result from bioinformation collection for the purposes of prosecutions where black and brown men are over-represented, given the widely documented racism in policing in the United States. Maguire and Murphy's chapter, on the other hand, is concerned with offering an ethnographic account of a key bioinformation processing site, the international airport. Forensic bioinformation gathering and processing at the airport illustrates the embeddedness of the medical/forensic interface in everyday life. It also draws attention to the interoperability of systems and the rhetorical and concrete force of analytics focused on databases integration. The chapters illustrate the endpoints of what we call the situated forensic bioinformation complex, that is, multifaceted bioinformation ecologies situated within national jurisdictions, but also transnational in scope. The situated forensic bioinformation complex rhetorically invokes visions of totality, is often extractive, but is also in practice open, overflowing and made out of gaps, blanks and bioinformational hiatuses.

In Chapter 6, 'Top_to_toe.ods: bioinformation and the politics of rape response', Sylvia McKelvie offers a compelling account of the vicissitudes of bioinformation collection in rape investigations in the United Kingdom. McKelvie shows how sexual assault survivors' personal data is captured by systematically invasive procedures that straddle the domains of medicine and criminal justice. McKelvie's contribution builds on feminist technoscience studies that have documented the material-semiotic practices of forensic science and rape response and shown how prototype kits are used to produce different types of knowledge and memories of assaults, to produce objective, neutral facts. However, a range of issues emerge as bodies are produced through these material-semiotic processes, throwing up quandaries for anti-rape movements. Here McKelvie deftly shows the limits of carceral feminisms (Bumiller 2008, Richie 2012) and their complicity with law and order agencies which are often gender blind and structurally racist, as they ground their quests for justice in the ability of bioinformation to produce identifications through a totalising bioinformation system. McKelvie argues that rather than an unproblematic route to justice in rape cases, the 'bodies as information' in sexual assault cases feed into national databases such the United Kingdom NDNAD and become a resource for state security (see biometrics), rather than gender justice claims. This requires new conversations between feminist technoscience and bioinformation studies and McKelvie outlines a compelling agenda for feminist approaches to bioinformation.

In Chapter 7, 'American bioinformation and U.S. race politics: the values of diverse genetic data', Anna Jabloner raises a range of compelling questions regarding how 'diversity' is figured across a range of bioinformation institutional and political imaginaries in the United States. Jabloner shows how 'diversity' is ascribed to data in different contexts, sustaining very different raciological (bio)political projects and biocapital calculations. The chapter illustrate how 'data that lack diversity' or 'super-diverse data' are not stable entities. Rather, over- and under-representation of some groups acquire differential meanings and value(s) across domains. Raciological orders are both reflected and constituted through the making of 'diversity' in bioinformation and Jabloner shows how 'diversity' is seen and unseen across forensic and health data ecologies in ways that do not align neatly. Diversity is invoked as a core value to underpin large research endeavours such as the All of Us Research Program, a venture which aims to capture diversity in terms of race, gender, sexuality and socio-economic status nationally. In forensic contexts such as the federal criminological genetic database CODIS, the over-representation of certain groups is not seen as a marker of diversity. Instead, here indexing is used to produce race neutrality, while at the same time obscuring the racism of policing and law enforcement institutions which produce the database in the first place. The over-representation of racially minoritised groups is offered as an argument for universal bioinformation gathering in biomedical research, targeting disproportionally black and brown men, arguably to redress the racism which results in 'diversity deficits'. It is in this way that the value of bioinformation data can be framed as universal. Jabloner contrasts these incommensurate claims about the value of bioinformation data on the grounds of diversity or universality, as articulated from situated, albeit unmarked, privileged vantage points. Building on Fortun's elegant rendition of the contradictory stakes and claims in the genomic biotech industry in Iceland through the figure of the chiasmus (2008), Jabloner shows the frictions that emerge in genomics in the context of American race politics along the nexus of specificity/universality and individuality/collectivity. The value of bioinformation changes across domains and infrastructures, as it transverses legal, biomedical, and consumer data ecologies.

Transnational and global bioinformation dynamics are the focus of Chapter 8. In a study of bioinformation regimes through practices of extraction and e-waste management, Peter Little raises key questions about the changing biopolitics of extraction and bioaccumulation. Little's analysis of e-waste management takes into account labour practices at facilities in China and Ghana, establishing a fruitful comparison between these cases to illuminate how bioinformation underpins forms of global biopolitics which demand that we think apparently different and contrasting practices in relation. Focusing on derivative information relating to toxicities and genomic instabilities, China and Ghana offer contrasting environmental health narratives that raise new questions about biopolitical critique, interrogating the politics of extraction, the value of aggregation and epidemiological metrics, and their connection with wider geographies of e-waste health management and expertise. Little follows environmental epidemiologists, e-waste workers, and shifting material forms of bioinformation to explore how local relations between labouring bodies and electronic discard policies expose patterns of e-waste toxicity connected to global bioinformation economies and the technoscientific systems that produce them. Little argues, following Lock (2017), that these wider historical and socio-political relations frame the significance of local bioinformation biologies in terms of a 'technoecobiopolitics' (Little 2012), a politics of life entangled in techno-environmental disruption, toxicity and contamination that becomes an object of regulation and expertise in global technoscience. Indeed, Little shows that bioinformation itself entangles these wider informational metanarratives in its constitution, management, and afterlives once it becomes part of particular research or policy assemblages.

In Chapter 9, 'Seeing like an airport: towards interoperability in contemporary security', Maguire and Murphy show, through a rich ethnographically grounded account, how the airport is a site of production of bioinformation in and through (bio)surveillance systems and infrastructures. The airport itself appears to be a multi-temporal site made out of modern bureaucratic practices that belong to the 19th and 20th centuries and that unfold alongside 21st-century biometrics and bioinformation-centred systems which seem to be increasingly interoperable, or at least promise increased integration, interoperability and efficacy. Maguire and Murphy approach the airport as a space of multi-temporal heterogeneity where modern biometrics and bioinformation surveillance systems unfold alongside each other, occasionally criss-crossing – and where remarkably stable social logics and taxonomies of population surveillance are reproduced. In this specific iteration, however, bioinformation at the airport is a material-semiotic object which is in part matter, part rhetoric. Interoperability is at once a rhetorical device that sustains the project of 'seeing like an airport' and a range of socially situated practices of surveillance. The future-oriented vision of interoperability and integration – a version of the 'total system' similar to the 'total archive' (Lemov 2015, Daston 2017) – however, is shown to entail in practice notable blank spots in the systems, exemplified by the corridors and pathways where 'the airport cannot see' and where, beyond the surveillance gaze, flora and fauna thrive. Maguire and Murphy's incisive account maps the blank spaces in the integrated systems model coded as the interoperable future. This has significant implications given that governments are currently investing heavily in operationalising automated predictive technologies which promise increased integration and interoperability. The UK strategy is the Home Office Biometric Strategy, for example, aims to combine fingerprints, DNA and facial images, in an integrative vision of post-archival data consolidation (Gonzalez-Polledo and Posocco, forthcoming). Open source expert systems promise to process ever-increasing volumes and types of data – biometrics, facial recognition, retina scans, and behavioural patterns alongside fingerprint capture, photo matching and a plethora of context-specific operations of data capture and processing. The analyses of bioinformation presented in the volume, however, considerably complicate the vision of totality, integration and optimised interpretation in expert and lay accounts.

In the final chapter, Kiheung Kim and Jongmi Kim focus on responses to the Covid-19 pandemic in South Korea in 2020. They show that the management of citizens' bioinformation was central to governmental strategies to contain the spread of the virus. However, they argue that to understand the readiness of South Koreans to adopt digital tracking methods such as tracing the movements of mobile phones and credit cards as well new technologies of testing in drive-thru and walk-thru methods, socio-historical factors should be considered. More specifically, the histories of bioinformation management in previous public health emergencies and histories of biocitizenship in the country are particularly salient. Kim and Kim relate the Covid-19 crisis to decades-long set of reforms that begun in earnest in the 1980s and which have progressively led to the construction and reframing of the disease controlling and preventing systems, through the responses to outbreaks of zoonotic infectious diseases including SARS, MERS, Foot-and-Mouth Disease and African Swine Fever. The chapter, then, connects developments in bioinformation management and integration to new forms of governmentality increasingly concerned with transparency and direct democratic processes in the aftermath of anti-government movements, the so-called Candlelight Movement, which led to the impeachment of former president Park Geun-hye. Unpacking Orientalist framings of South Koreans as docile citizens unable or unwilling to contrast governmental intrusion into everyday life due to Confucian traditions and ethics, the chapter presents bioinformation in the age of Covid-19 as a postcolonial artefact which diffracts socially situated anxieties about the status of power and the parameters of agency at play in voluntary bioinformation relay and surrender. As Covid-19 recast yet again questions about bioinformation governance and infrastructures with new urgency and in a new light, the case of South Korea is an important reminder of the way bioinformation histories can at once sediment and emerge in the ordinariness of life in the pandemic.

References

Agamben, Giorgio. 1998. *Homo Sacer: Sovereign Power and Bare Life*. Stanford: Stanford University Press.

Amoore, L. 2018: 'Cloud Geographies: Computing, Data, Sovereignty'. *Progress in Human Geography*, 42(1): 4–24.

Anderson, Chris. 2008. 'The End of Theory: The Data Deluge Makes the Scientific Method Obsolete'. *Wired*, 23 June 2008. Available at: www.wired.com/science/discoveries/magazine/16-07/pb_theory.

Bateson, Gregory. 1979. *Mind and Nature: A Necessary Unity*. New York: E.P. Dutton.

Biehl, Joao. 2001. *Vita: Life in a Zone of Social Abandonment*. Los Angeles: University of California Press.

Bumiller, Kristin. 2008. *In an Abusive State: How Neoliberalism Appropriated the Feminist Movement against Sexual Violence*. Durham, NC: Duke University Press.

Cassidy, Rebecca. 2002. *The Sport of Kings: Kinship, Class and Thoroughbred Breeding in Newmarket*. Cambridge: Cambridge University Press.

Comaroff, Jean. 2007. 'Beyond the Politics of Bare Life: AIDS and the Global Order'. *Public Culture*, 19(1): 197–219.

Daston, Lorraine. 2017. *Science in the Archives: Pasts, Presents, Futures*. Chicago & London: University of Chicago Press.

Farmer, Paul. 1996. 'On Suffering and Structural Violence: A View from Below'. *Daedalus*, 125(1): 261–283.

Franklin, Sarah. 2007. *Dolly Mixtures: The Remaking of Genealogy*. Durham, NC: Duke University Press.

Gibbon, Sahra and Novas, Carlos. 2008. *Biosocialities, Genetics and the Social Sciences: Making Biologies and Identities*. London: Routledge.

Gonzalez-Polledo, EJ and Posocco, Silvia. Forthcoming. 'Forensic Apophenia: Sensing the Bioinformation Archive'. *Anthropological Quarterly*.

Medina, Eden. 2011. *Cybernetic Revolutionaries: Technology and Politics in Allende's Chile*. Cambridge, MA: MIT Press.

Ferretti, Luca, Chris Wymant, Michelle Kendall, Lele Zhao, Anel Nurtay, Lucie Abeler-Dörner, Michael Parker, David Bonsall, and Christophe Fraser. 2020. 'Quantifying SARS-CoV-2 Transmission Suggests Epidemic Control with Digital Contact Tracing'. *Science*, 368(6491). doi:10.1126/science.abb6936.

Forster, Peter, Lucy Forster, Colin Renfrew, and Michael Forster. 2020. 'Phylogenetic Network Analysis of SARS-CoV-2 Genomes'. *Proceedings of the National Academy of Sciences*, April. doi:10.1073/pnas.2004999117.

Foucault, Michel. 1981. *The History of Sexuality Vol. I*. London: Penguin.

Foucault, Michel. 2001. 'Governmentality', in Faubian, James D. (Ed), *Power: Essential Works of Foucault 1954–1989*. London: Allan Lane.

Foucault, Michel. 2003. *'Society Must be Defended': Lectures at the Collège de France, 1975–1976*. London: St Martin's Press.

Fortun, Mike. 2019. 'Six Impossible Things About GeneXEnvironment Interactions: Care of Data, Selves, and Systems in Genomics and Air Pollution Research', Birkbeck, University of London, 21 June 2019.

Fortun, Mike. 2008. *Promising Genomics: Iceland and DeCODE Genetics in a World of Speculation*. Berkeley, CA: University of California Press.

Foucault, Michel. 1991. *Discipline and Punish: The Birth of the Prison*. London: Penguin Books.

Franklin, Sarah. 2003. 'Re-thinking Nature–Culture: Anthropology and the New Genetics'. *Anthropological Theory*, 3(1): 65–85.

Haraway, Donna J. 1988. 'Situated Knowledges: The Science Question in Feminism and the Privilege of Partial Perspective', *Feminist Studies*, 14(3):575–599.

Harvey, Penny and Knox, Hannah. 2015. *Roads: An Anthropology of Infrastructure and Expertise*. Ithaca, New York: Cornell University Press.

Harvey, P., Jensen, C. B, & Morita, A. (Eds.) 2017. *Infrastructures and Social Complexity: A Companion*. London: Routledge.

Hirsch, Eric, and Marilyn Strathern (Eds.) 2006. *Transactions and Creations: Property Debates and the Stimulus of Melanesia*. New York: Berghahn Books.

Hoeyer, Klaus. 2013. *Exchanging Human Bodily Material: Rethinking Bodies and Markets*. Amsterdam: Springer. https://doi.org/10.1007/978-94-007-5264-1.

Irani, Lilly. 2019. *Chasing Innovation: Making Entrepreneurial Citizens in Modern India*. Princeton Studies in Culture and Technology. Princeton & Oxford: Princeton University Press.

Kay, Lily E. 2000. *Who Wrote the Book of Life?: A History of the Genetic Code*. Writing Science. Stanford, CA: Stanford University Press.

Kennedy, Helen. 2016. *Post, Mine, Repeat: Social Media Data Mining Becomes Ordinary*. London: Palgrave Macmillan.

Koopman, Colin. 2019. *How We Became Our Data: A Genealogy of the Informational Person*. Chicago: The University of Chicago Press.

Lemov, Rebecca M. 2015. *Database of Dreams: The Lost Quest to Catalog Humanity*. New Haven & London: Yale University Press.

Leonelli, Sabina. 2016. *Data-Centric Biology: A Philosophical Study*. Chicago & London: The University of Chicago Press.

Lindtner, Silvia M. 2020. *Prototype Nation: China and the Contested Promise of Innovation*. http://public.eblib.com/choice/PublicFullRecord.aspx?p=6260074.

Little, Peter C. 2012. *Think Technoecobiopolitics: Reflecting on the Emerging Political Ecology of IBM's 'Smarter Planet' Mission*. Paper presented at the American Anthropological Association Annual Meeting, San Francisco, USA, November.

Lock, Margaret. 2017. 'Recovering the Body'. *Annual Review of Anthropology*, 46:1–14.

Marres, Noortje, and David Stark. 2020. 'Put to the Test: For a New Sociology of Testing'. *The British Journal of Sociology*. doi:10.1111/1468-4446.12746.

Mayer-Schönberger, Viktor, and Kenneth Cukier. 2013. *Big Data: A Revolution That Will Transform How We Live, Work, and Think*. Boston: Houghton Mifflin Harcourt.

M'charek, Amade. 2020. 'Tentacular Faces: Race and the Return of the Phenotype in Forensic Identification'. *American Anthropologist*, 122(2): 369–380. doi:10.1111/aman.13385.

McMullin, Barry. 2000. 'John von Neumann and the Evolutionary Growth of Complexity: Looking Backward, Looking Forward'. *Artificial Life*, 6(4): 347–361. doi:10.1162/106454600300103674.

Medina, Eden. 2011. *Cybernetic Revolutionaries: Technology and Politics in Allende's Chile*. Cambridge, MA: MIT Press.

Mitman, Gregg, Murphy, Michelle and Sellers, Christopher (Eds). 2004. *Landscapes of Exposure: Knowledge and Illness in Modern Environments*. Durham & London: Duke University Press.

Nisa, Richard. 2016. 'Bodies of Information: Data, Distance and Decision-Making at the Limits of the War Prison'. In *Algorithmic Life: Calculative Devices in the Age of Big Data*, edited by Louise Amoore. London & New York: Routledge.

Oyama, Susan. 2000. *The Ontogeny of Information: Developmental Systems and Evolution*. Durham & London: Duke University Press.

Parreñas, Juno Salazar. 2018. *Decolonizing Extinction: The Work of Care in Orangutan Rehabilitation*. Durham & London: Duke University Press.

Pálsson, G. (2007). *Anthropology and the New Genetics*. Cambridge: Cambridge University Press.

Pálsson, G. (2008), 'Genomic Anthropology: Coming in from the Cold?' *Current Anthropology*, 49 (4): 545–568.

Parry, Bronwyn. 2004. *Trading the Genome: Investigating the Commodification of Bio-Information*. New York: Columbia University Press.

Parry, Bronwyn, and Beth Greenhough. 2018. *Bioinformation*. Cambridge: Polity.

Petryna, Adriana. 2013. *Life Exposed: Biological Citizens after Chernobyl*. Princeton, NJ: Princeton University Press.

Povinelli, Elizabeth. 2011. *Economies of Abandonment: Social Belonging and Endurance in Late Liberalism*. Durham & London: Duke University Press.

Prainsack, Barbara, Schicktanz, Silke and Werner-Felmayer, Gabriele. 2014. *Genetics as Social Practice: Transdisciplinary Views on Science and Culture*. London & New York, Routledge.

Rabinow, Paul. 1996. 'Artificiality and Enlightenment: From Sociobiology to Biosociality'. In *Essays on the Anthropology of Reason*. Princeton: Princeton University Press.

Rabinow, Paul and Rose, Nikolas. 2006. 'Biopower Today'. *BioSocieties*, 1: 195–217.

Regalado, A. 2019. 'More Than 26 Million People Have Taken an At-home Ancestry Test'. *MIT Technology Review*, February 11, 2019.

Richardson, Sarah S. 2013. *Sex Itself: The Search for Male and Female in the Human Genome*. Chicago: Chicago University Press.

Richie, Beth E. 2012. *Arrested Justice: Black Women, Violence, and America's Prison Nation*. New York: New York University Press.

Riles, Annelise. 1998. 'Infinity Within the Brackets'. *American Ethnologist*, 25(3): 378–398.

Rose, Nikolas. 2001. 'The Politics of Life Itself'. *Theory, Culture and Society*, 18(6): 1–30.

Rose, Nikolas. 2007. 'Molecular Biopolitics, Somatic Ethics and the Spirit of Biocapital'. *Social Theory & Health*, 5(1): 3–29. doi:10.1057/palgrave.sth.8700084.

Rose, Nikolas, and Carlos Novas. 2005. 'Biological Citizenship'. In *Global Assemblages*, edited by Aihwa Ong and Stephen J. Collier, 439–463. Oxford: Blackwell. doi:10.1002/9780470696569.ch23.

Shaw, Alison. 2009. *Negotiating Risk: British Pakistani Experiences of Genetics*. New York: Berghahn Books.

Strathern, Marilyn. 2020. *Relations: An Anthropological Account*. Durham & London: Duke University Press.

TallBear, Kim. 2013. *Native American DNA: Tribal Belonging and the False Promise of Genetic Science*. Minneapolis: Minnesota University Press.

Tamarkin, Noah. 2014. 'Genetic Diaspora: Producing Knowledge of Genes and Jews in Rural South Africa'. *Cultural Anthropology*, 29(3): 552–574.

Taussig, Karen Sue. 2009. *Ordinary Genomes: Science, Citizenship, and Genetic Identities*. Durham & London: Duke University Press.

Thompson, Charis. 2005. *Making Parents: The Ontological Choreography of Reproductive Technologies*. Cambridge, MA: MIT Press.

Turkle, Sherry. 2009. *Simulation and Its Discontents*. Cambridge, MA: MIT Press.

Ventura Santos, Ricardo and Wade, Peter. 2014. 'Negotiating Imagined Genetic Communities: Unity and Diversity in Brazilian Science and Society'. *American Anthropologist*, 116(4): 736–748.

2   All the data creatures

Tahani Nadim

CACCTTATACTTTATCTTCGGCGCATGAGCTGGAATAGTGGGAACA
GCCCTAAGCCTTCTAATCCGGACAGAACTAAGCCAACCTGGCCCTC
TTATAGGAGATGATCAGATTTACAATGTCATTGTCACAGCACATGC
CTTTATTATAATTTTCTTTATGGTAATACCAATCATAATCGGAGGA
TTTGGAAACTGGCTACTGCCACTAATAATCGGAGCACCCGATATAG
CATTTCCCCGAATAAATAACATAAGCTTCTGATTATTACCGCCCTC
ATTTATCTTACTTCTATTTTCCGCCTTCATTGAAACAGGCGCCGGT
ACAGGATGAACTGTTTACCCACCACTAGCTGGGAATCTGGCCCACG
CTGGGCCATCAGTAGATTTAACTATCTTCTCCCTTCATCTTGCTGG
GGTTTCATCAATTTTAGGGGCAATCAACTTTATCACCACAGCCCTT
AATATAAAACCACCATCGATGTCACAACAACAAACACCACTTTTCG
TATGATCCGTACTAGTCACGGCTGTACTCCTACTGCTCGCCCTACC
GGTCCTAGCGGCAGGAATTACTATATTACTCACTGACCGAAACTTA
AACACTACCTTTTTCGATCCTGCTGGAGGAGGGGATCCAATCCTATA

The textual form a DNA sequence takes is familiar. It is used as stylistic element in cover designs and film titles, running across screens and devices. It is one of the most recognizable data products of the biosciences, one of the many millions of sequences produced in genome centres and laboratories every day. It embodies a complex signifier that flickers between natural and cultural categories, signalling a diverse, at times incongruous set of hopes, futures and fears. DNA sequence, the order of the four chemical bases – cytosine, adenine, guanine, thymine – that make up DNA molecules that make up genes and genomes, also functions as a critical protagonist in the identification of animal species. The sequence printed above represents part of the genome of a West African dwarf crocodile. More specifically, it represents part of the COX1 gene, the so-called "barcode", a standardized short sequence of DNA that – in principle – should characterize all animal species on the planet (Savolainen et al. 2005). The barcode thus functions as the genetic signature for the West African dwarf crocodile, *Osteolaemus tetraspis*, which was first described and named in 1860 in the *Proceedings of the Academy of Natural Sciences of Philadelphia*.

Recently, debate has emerged about the legal definition of "digital sequence information" (DSI), a relatively new term referring to the informational

DOI: 10.4324/9780367810030-2

potentials residing within organisms. Discussions first appeared in the context of the Convention on Biological Diversity (CBD), specifically in relation to access and benefit-sharing agreements meant to ensure a fair and equitable share in the benefits of scientific discoveries for providers and holders of biological materials. But discussions can also be tracked in the International Treaty on Plant Genetic Resources for Food and Agriculture, the WHO Pandemic Influenza Preparedness Framework as well as the UN Convention on the Law of the Sea. At the heart of the debate lie hundreds of years of extraction and exploitation of lands and peoples and the persistent inequalities and ruinations they cause. However, these are concentrated into one key issue: the nature of sequence data. In a 2017 submission to the Secretariat of the Convention on Biological Diversity, the African Centre for Biodiversity, a research and advocacy organization working towards food sovereignty and agroecology in Africa, makes the case to "unequivocally require that sequence data be considered equivalent to its physical biological counterparts". The argument suggests that the barcode should be regarded as equivalent to the crocodile, its tissues and samples used in the extraction of DNA and in the production of the barcode. Unlike sequence data, biological materials are governed more stringently under access and benefit-sharing agreements and other legal instruments regulating the collection, export and distribution of biological materials.

In light of these debates, I want to examine how barcode sequences and their material biological counterparts inform each other. In this chapter I am therefore concerned with the historical materialities and continuities of the barcode sequence which thus becomes part of larger "data formations" that connect practices and institutions across times and places (Nadim 2021). Specifically, I engage sequence data as an empirical terrain through which the value and concept of species are negotiated, enacted and contested. Indeed, the question of how to locate and access – not just in the sense of using it but of thinking with it – DNA sequence data is as important to social and cultural enquiry as it is to the communities that produce, collect, maintain and use this data. Figuring the place of data is instructive for provincializing the claims and methods of data-driven sciences, that is, for historicizing and situating data-based self-evidences. Where data is – and the related question of when data is – constitute a productive approach for understanding the stakes involved in struggles over the nature of data.

The chapter first attends to the technology of barcoding and its politics, which I situate within a colonial desire to collect, name and classify species. I then describe and discuss three "bioinformational artefacts" (Nadim 2012): a database record from the genetic sequence database GenBank, a page from a catalogue recording accessions to the Museum für Naturkunde Berlin (the natural history museum in Berlin), and the scientific paper that has given rise to the barcode printed above. I conclude by discussing the informational constitution of species as data creatures and highlighting the mutual implication of "race" and "species" as a critical aspect to consider in debates on the nature of digital sequence information. My analysis draws from an ongoing project based at the museum, a practice-focused ethnographic study of data formations in and with natural history. This is informed by science and technology studies and the history of science and is concerned with the role of data in the traffic between social and biological life. My central method has been that of changing between observer and participant. I have conducted interviews and workplace observations at GenBank and the European Nucleotide Archive in 2008 and, since 2013, at the Museum where I head the department Humanities of Nature and participated, among other things, in taxonomic research, the digitization project and grant-writing efforts around developing data infrastructures (ongoing).

Unpacking barcoding

The barcode method was developed in the early 2000s by researchers at the University of Guelph in Canada to assist in the identification of animal species (Hebert et al. 2003, Hebert and Gregory 2005). The barcoding method requires the building of barcode libraries of known species and the assigning of barcodes to taxonomically unknown specimens using matching algorithms. It quickly spread across research communities and by the end of the decade had generated over four million barcodes for 500,000 species of animals, plants and fungi. Enormous digital archives of barcodes, such as the Barcode of Life database (BOLD), serve as the standard against which all DNA barcode sequences generated from unknown samples can be matched (Ratnasingham and Hebert 2007). In addition, the promise of the barcode is driving ever grander initiatives such as the Earth BioGenome Project, which was officially launched at the World Economic Forum in Davos in 2018, and intends to generate and make publicly available not just barcodes – which are part of genomes – but entire genome data for all known species of animals, plants, and fungi. In order to get at the still outstanding 99.8% of all unknown eukaryotic species, the project involves, among other things, the construction of new DNA and tissue biobanks, surveys and sample collection of biodiversity from a minimum of five biodiversity hotspots such as Brazil and Madagascar, as well as a series of legal agreements. It will comprise a standardized and streamlined bioinformational pipeline, so a set of computational and – where possible – automated routines through which genomes get assembled and species identified. Sample acquisition and identification might even include the use of "aerial, terrestrial, and aquatic autonomous drones equipped with high-resolution cameras that can enable species collection and identification and telecommunications with taxonomic experts" (Lewin et al. 2018: 4330).

Biologists, ecologists, policy-makers and organizations dedicated to wildlife monitoring are advocating for genetic barcodes as more efficient, reliable and, importantly, doable means for species identification. Traditionally, identification is based on the morphological examination of specimens and the comparison of morphological traits by taxonomists. Classical taxonomy requires time: obtaining and preparing specimens, collating and reviewing literature on already identified species, careful examination, comparison and description, and, finally, publication of the description in a recognized outlet. It also requires highly specialized expertise given the extent of biodiversity and the smallness of difference. Many experts in conservation biology and policy refer to this as the "taxonomic impediment": the slowness of traditional methods for species identification compounds biodiversity loss and destruction since these can only be addressed once there is a name for what's there and a centralized inventory for managing its data. "Can we name Earth's species before they go extinct" (Costello et al. 2013) has been a rallying cry for proponents of semi-automated solutions such as barcoding. To barcoding advocates, the future of biodiversity and conservation is data-based because naturally everything can and, importantly, *should* be captured in and with data.

Barcoding histories

I am highlighting the normative dimension of this effort not only because it is so readily found in the barcoding literature, which habitually stresses the urgency and moral imperative of knowing species. But honing in on the ways in which barcoding and, more generally, the datafication of nature become seen as commonsensical opens an analytical focus for drawing out historical continuities across practices dedicated to collecting and ordering nature. The Bio-Genome Project is very explicit about its dependence on the involvement of and coordination with all major natural history museums and collections as they house the basic comparative stock of specimens, the physical biological counterparts to barcodes and DSI, as well as many yet-to-be identified specimens (that could potentially turn out to be new species). Historical continuities subtend the drive toward data-intensive biodiversity identification and point to the complex temporalities enfolded within data and data practices. Current large-scale data-gathering ventures seeking to produce total archives of species are, I would contend, steeped in similar epistemic habits that condition the realm of the doable and desirable.

The desire for total knowledge has been a key trope in the pursuit of natural history and its unceasing attempts to create inventories of all life on Earth. It is palpable in the collections of natural history museums, including the Museum für Naturkunde Berlin, which teem with millions and millions of dead animals, killed and collected to satisfy a desire to have one (or two or five or 67) of every kind. It is also discernable in the shape of recording technologies such as lists and ledgers which are infinitely expandable and thus spell no end to the collection of nature. Transforming the accumulative impulse into a reasonable, commonsensical objective is intimately tied to the idea of a total archive pursued by colonial empires (Richards 1993). The colonial state was a rapacious data production machine, intent on, as Richards put it, "superintending all knowledge, particularly the great reams of knowledge coming in from all parts of the Empire" (Richards 1993: 6). Data was regarded as a prerequisite for control at a distance and it was a means by which to enact the violent objectification of colonized peoples and lands while giving the host of colonial administrators, officials and personnel something to do. In her work on the colonial archives of the Dutch Indies (present-day Indonesia), the historical anthropologist Ann Laura Stoler suggests that "[c]olonial statecraft was an administrative apparatus" producing "weekly reports to superiors, summaries of reports of reports, and recommendations based on reports" compelling the establishment of complex (information) infrastructures (Stoler 2009: 29). This was also the time when museum collections all over the Western metropolises filled with animals and artefacts, prompting, as Foucault so vividly described in *The Order of Things*, new forms (such as Linnaean taxonomy) and settings (such as museums) for perceiving them (1970).

But spending time inside museum collections quickly disabuses any visitor of such totalizing data visions. While there are indeed lots and lots of artefacts, objects, and specimens as well as an abundance of documentation in various forms, there is no overarching order, no ultimate logic, no scientific theory that would turn this accumulation into a sensible, intelligible unit. A similar revelation has greeted science studies scholars (e.g. Knorr-Cetina 1999) upon entering the laboratories of science, which did not correspond to the clearly and cleanly structured arguments made in the scientific papers. This prompted scholars to attend to how knowledge claims were negotiated in the messy practices of laboratory life and how they then travelled outside the laboratory walls unencumbered, "purified," as Latour put it, of their creative and tentative origins. In analysing colonial statecraft, Stoler draws parallels to the epistemic practices of the sciences. Examining colonial governing practices – and museum collections were both products of and instruments for these practices – makes clear how knowledge did not automatically equate with power and that the panoptic gaze worked more as a rhetorical mode than a technical function of the colonial archive.

I find these observations instructive not just for disenchanting the claims about the prowess of barcoding and genetic data (and the authority of museum collections) but because they compel a closer look at situated knowledge and its "epistemic habits". The latter refers to "ways of knowing that are available and 'easy to think,' called-upon, temporarily settled dispositions that can be challenged and that change" (Stoler 2009: 39). In her study, Stoler closely reads archival documents for the traces of these habits and the marks of sentiments that would reveal the minor and major attachments constitutive of wider colonial rule. In the following I am guided by this methodological approach in describing and discussing three "bioinformational artefacts", a database record, a page from a museum register and a scientific paper (Nadim 2012). I define bioinformational artefacts as both products and subjects of archives, databases and other efforts dedicated to cataloguing biological life. They are at once physical and informational, connecting social and natural orders. And lastly, as artefacts of modern knowledge practices, they index historical and ongoing agreements (and disagreements) over the nature of data and the data of nature.

Bioinformational artefact I: the GenBank record

Figure 2.1 (all images at the end) shows the complete GenBank database record for the DNA barcode sequence shown at the beginning of this paper. The GenBank database is produced and maintained by the National Center of Biotechnology Information (NCBI), a division of the National Library of Medicine

Osteolaemus tetraspis tetraspis voucher M11 cytochrome oxi... https://www.ncbi.nlm.nih.gov/nuccore/GQ144621.1?report=...

Nucleotide

The Nucleotide database will include EST and GSS sequences in early 2019. Read more

GenBank (full)

Osteolaemus tetraspis tetraspis voucher M11 cytochrome oxidase subunit 1 (COI) gene, partial cds; mitochondrial

GenBank: GQ144621.1

FASTA Graphics

Go to:

```
LOCUS       GQ144621                 645 bp    DNA     linear   VRT 24-JUL-2016
DEFINITION  Osteolaemus tetraspis tetraspis voucher M11 cytochrome oxidase
            subunit 1 (COI) gene, partial cds; mitochondrial.
ACCESSION   GQ144621
VERSION     GQ144621.1
KEYWORDS    BARCODE.
SOURCE      mitochondrion Osteolaemus tetraspis tetraspis
  ORGANISM  Osteolaemus tetraspis tetraspis
            Eukaryota; Metazoa; Chordata; Craniata; Vertebrata; Euteleostomi;
            Archelosauria; Archosauria; Crocodylia; Longirostres; Crocodylidae;
            Osteolaemus.
REFERENCE   1  (bases 1 to 645)
  AUTHORS   Eaton,M.J., Meyers,G.L., Kolokotronis,S.-O., Leslie,M.S.,
            Martin,A.P. and Amato,G.
  TITLE     Barcoding bushmeat: molecular identification of Central African and
            South American harvested vertebrates
  JOURNAL   Conserv. Genet. 11 (4), 1389-1404 (2010)
REFERENCE   2  (bases 1 to 645)
  AUTHORS   Eaton,M.J., Meyers,G.L., Kolokotronis,S.-O., Leslie,M.S.,
            Martin,A.P. and Amato,G.
  TITLE     Direct Submission
  JOURNAL   Submitted (12-MAY-2009) Patuxent Wildlife Research Center, United
            States Geological Survey, 12100 Beech Forest Road, Laurel, MD
            20708-4017, USA
FEATURES             Location/Qualifiers
     source          1..645
                     /organism="Osteolaemus tetraspis tetraspis"
                     /organelle="mitochondrion"
                     /mol_type="genomic DNA"
                     /sub_species="tetraspis"
                     /specimen_voucher="M11"
                     /db_xref="taxon:184243"
                     /country="Cameroon"
                     /PCR_primers="fwd_seq:
                     tgtaaaacgacggccagtcaaccaaccacaaagacattggcac, rev_seq:
                     caggaaacagctatgacacttcagggtgaccgaagaatcagaa"
                     /PCR_primers="fwd_seq:
                     tgtaaaacgacggccagtcgactaatcataaagatatcggcac, rev_seq:
                     caggaaacagctatgacacctcagggtgtccgaaraaycaraa"
     gene            <1..>645
                     /gene="COI"
     CDS             <1..>645
                     /gene="COI"
                     /codon_start=2
                     /transl_table=2
                     /product="cytochrome oxidase subunit 1"
                     /protein_id="ACS44998.1"
                     /translation="TLYFIFGAWAGMVGTALSLLIRTELSQPGPLMGDDQIYNVIVTA
                     HAFIMIFFMVMPIMIGGFGNWLLPLMIGAPDMAFPRMNNMSFWLLPPSFILLLFSAFI
                     ETGAGTGWTVYPPLAGNLAHAGPSVDLTIFSLHLAGVSSILGAINFITTALNMKPPSM
                     SQQQTPLFVWSVLVTAVLLLLALPVLAAGITMLLTDRNLNTTFFDPAGGGDPIL"
ORIGIN
        1 caccttatac tttatcttcg gcgcatgagc tggaatagtg ggaacagccc taagccttct
       61 aatccggaca gaactaagcc aacctggccc tcttatagga gatgatcaga tttacaatgt
      121 cattgtcaca gcacatgcct ttattataat tttctttatg gtaataccaa tcataatcgg
      181 aggatttgga aactggctac tgccactaat aatcggagca cccgatatag catttccccg
      241 aataaataac ataagcttct gattattacc gccctcattt atcttacttc tattttccgc
      301 cttcattgaa acaggcgccg gtacaggatg aactgtttac ccacccactag ctgggaatct
      361 ggcccacgct gggccatcag tagatttaac tatcttctcc cttcatcttg ctggggtttc
      421 atcaatttta ggggcaatca actttatcac cacagccctt aatataaaac caccatcgat
      481 gtcacaacaa caaacaccac ttttcgtatg atccgtacta gtcacggctg tactcctact
      541 gctcgcccta ccggtcctag cggcaggaat tactatatta ctcactgacc gaaacttaaa
      601 cactaccttt ttcgatcctg ctggaggagg ggatccaatc ctata
//
```

Figure 2.1 GenBank database record
Source: www.ncbi.nlm.nih.gov/nuccore/GQ144621.1.

and part of the National Institutes of Health. It is the largest public archive for DNA and RNA sequence data and forms a single repository together with the European Nucleotide Archive (ENA) and the DNA Database of Japan (they exchange their data daily). Together, they are referred to as the International Nucleotide Sequence Database Collaboration (INSDC). Sequence data in these archives is derived from all kinds of organisms, viruses as well as synthetic sequences. GenBank is therefore an essential repository and research tool for the biosciences and an obligatory passage point for many researchers since journals mandate deposit of sequences generated as part of a publication in one of the three databases. GenBank currently contains more than 200,000,000 sequence records and is growing daily. Each record accounts for a single contiguous DNA or RNA sequence that is contextualized with various data points, summarily called annotation. This provides metadata, data about data, and is checked and often completed or refined by so-called curators, scientists – most educated to PhD level – that ensure the quality and integrity of submitted data. They are based in the GenBank offices in Bethesda, Maryland and form, together with taxonomists, engineers, information scientists and software developers, the major human element of the database (Nadim 2016).

The sequence appears at the very bottom of the record and is prepended by a corollary of contextual data that helps situate and qualify the sequence. Such data about data is crucial for data's portability and valence. Metadata makes up data's "packaging" (Leonelli 2010) and allows it to be linked and put in relation with other sequences (and species), domains, practices and institutions. The work of metadata, to Leonelli, eases the "tension between the local nature of facts about organisms and the need for them to circulate across widely different research contexts and locations" (Leonelli 2010: 326). Metadata thus plays a crucial role in what the editors have called the "legibility" and "the global reach of bioinformation infrastructures" (Introduction, this volume) because it furnishes data with the capacity to relate to other data. At the same time, metadata is meant to retain the original context of data extraction and production, which is not only important for guaranteeing validity but for guiding proper re-use and interpretation. It is important for researchers wanting to reuse the data to know where and how the sequence was generated in the first place, which instruments and materials were used and what the bioinformational pipeline looked like, the chain of algorithms and models used to construct DNA sequences from the chromatograms or electropherograms (graphical representations based on photographic images) produced by sequencing machines.

Looking at the record it is not obvious what constitutes or distinguishes data and metadata. Indeed, it is not obvious what's going on at all in this document where familiar words, expert language, cryptic abbreviations, codes, sequences and numbers accumulate. To be sure, these database records are not necessarily meant to be read by humans. Instead, they are there to be processed by algorithms as part of queries run on the vast archive of GenBank data. But by considering them *bioinformational artefacts*, they compel description, deduction and speculation for they evidence human-data mediations and attune us – critical scholars of bioinformation – to how data-based habits tangle with data-driven logics. For example, the contents and configurations of data and metadata in the GenBank record point to different types of origin.

Importantly, in the case of the crocodile's barcode sequence, the metadata in the database record ensures that the barcode can be traced back to a specific crocodile. A reference to the physical specimen "behind" the sequence already features in the title, "Osteolaemus tetraspis tetraspis voucher M11 cytochrome oxidase subunit 1 (COI) gene, partial cds; mitochondrial." "Voucher M11" refers to the sample's so-called "voucher specimen", the physical source for the DNA. Voucher specimens refer to preserved specimens used as supporting evidence for identification as well as for data generated in the course of a particular study. They are usually stored in museum collections and herbaria, representing a permanent record for the material used in such study. Examining the "Source" category for details about the physical referent of the sequence data, we learn, among other things, the species name, the specimen's voucher code and the country of origin, Cameroon. From the scientific paper referenced on the record, we learn that M11 resides in the American Museum of Natural History but their catalogue only records the type of preparation ("dry"), the part prepared ("skeletal, cranial") and the country of origin as "Cameroons", the spelling of which would indicate a collection date before independence in 1960 and 1961, respectively.

Also contained within the category is a link to the NCBI Taxonomy Database, a data resource that contains curated classification and nomenclature for all organisms in the public sequence databases. Following this link, we are given, among other things, two further links to external taxonomic information sources. The first links to the Encyclopedia of Life (EoL), maintained by the Smithsonian's National Museum of Natural History (Washington, D.C.), where a photograph of a whole dry crocodile specimen greets the visitor. The crocodile is splayed out, a tape measure is set slightly above its middle section and a big yellowed label dangles from its front left leg. The image caption bears the name of the Museum of Comparative Zoology (MCZ) at Harvard University. This maintains its own collection database, the MCZBase, which holds a very comprehensive record for this crocodile. But it is not the specimen "M11" which the data in the barcode record is derived from. Instead, the verbatim information from the specimen's label tells us that this crocodile had been caught in "Niapu, NE Belgian Congo" as part of what the MCZBase record identifies as the "Lang-Chapin Expedition 1909–1915". Sponsored by the American Natural History Museum under the enthusiastic approval of Belgium's King Leopold, this "expedition" was led by the German-born zoologist Herbert Lang and ornithologist James Chapin. It was, as historian Jeannette E. Jones writes, "an experiment in both zoological and anthropological knowledge production" which turned the Congo into an object of scientific enquiry and "model for American naturalist discourses on Africa" that welded together the workings of race and nature in particular ways (2011: 71–2).

Historians of (colonial) science have shown how the collection of nature in the course of colonial "expeditions" went hand in hand with the appropriation of cultural artefacts and the production of ethnographic accounts seeking to validate racial classifications. Unravelling the database record, this entangled history becomes evident through a yellowed paper label hanging from a dead crocodile. It does not require much digging to encounter histories of violence that subtend the collection infrastructures of modern knowledge practices. Genetic barcodes, while promising a liberation from the method of classical taxonomy, cannot escape the fact that their validity and re-usability remain tied to these material infrastructures. This relation can become visible through their voucher specimens as well as through their connection to Linnaean taxonomy expressed in species names and their placement within a hierarchical order of life based on types. Yet, genetic barcodes read as bioinformational artefacts might also point toward a less naturalized concept of species difference: focusing on molecular differences as the key for drawing boundaries around species has shifted attention to the ways in which institutions such as museums make available their objects and data because they are the ones guaranteeing the validity of barcodes. The database record proliferates relations. Instead of a single originary referent, links and histories appear.

Bioinformational artefact II: the *Eingangsbuch*

Administering collections compels diligent and relentless recording and this, I argue, is not a dispassionate labour. Specimens in collections of natural history museums are customarily entangled with inscriptions bearing data points. These include handwritten labels like the one hanging from the dead crocodile as well as labels on drawers and cupboards and, increasingly, QR codes. This data physically stays with the specimen wherever it might travel (and specimens do travel within the museum as well as beyond the museum as part of a lively global loan traffic between collections). It records key metadata including species name, the location of where it was taken from and the collection date. The label makes the specimen. Without it, it would only be a dead animal or plant or a random rock. Museum collections and seed banks (and other collections) form the material backbone for the dominating order of life on Earth and it is all held together by slips of paper. This dependency on fragile and often hard to decipher pieces of paper continues to amaze and rattle me, also because it remains so central in ongoing collection-based data practices (in the form of, for example, thousands of newly generated paper labels bearing tiny QR codes). The millions of handwritten and computer-printed labels express a "phantasmal belief" (Vismann 2016: 97) that data can contain the diversity of organisms and that data standards and procedures can vanquish externalities, such as colonial histories, or inaccuracies, such as outdated species names (the West African dwarf crocodile has, according to the Reptile Database,[1] 17 synonyms). Another textual layer, which exists at a physical distance from the actual specimen, can be found in the various written records that record its arrival, accession and, where appropriate, departure. It is in these pages that the moral values and attachments which shape and are shaped by data can be seen most clearly.

Figure 2.2 shows a double page from an *Eingangsbuch*, the register or, literally, the book of arrivals. It is kept in the offices of the curatorial staff for the Herpetology Department, the department for reptiles and amphibians at the Museum for Natural History Berlin. The *Eingangsbuch* is a leather-bound volume, part of a series that begins in the mid-1850s and runs into the 1940s. The book of arrival records incoming specimens, in this case, specimens that had arrived between April and August 1914. Many such books populate museum collections worldwide. Specimens might have been collected as part of Museum-sponsored surveys. They might have been donated by scientists, colonial officers, collectors and other institutions such as the Berlin Zoo. Or they might have been acquired from commercial traders specializing in supplying animals – dead or alive. This provenance – the nature of the transaction by which dead animals came into collections – is not usually found on the labels but noted in these registers. The form of the *Eingangsbuch* mirrors a double-entry bookkeeping ledger: there is a column for income (*Zugang*) and one for departure (*Abgang*), each followed by columns for recording their type and date. The full set of columns comprises the following data points: number, quantity of specimen, description, the location or find spot, date and type of accession, monetary value, date and type of de-accession, and notes.

The Post-it notes stuck over entry Nos. 692 and 698 as well as poking from other pages of the book show that this volume has been worked on and remains part of current curatorial practices. In the second row under the heading *Abgang* (outgoing), a number is written in pencil in a different script. Indeed, annotations from different hands and times animate the many written lists as well as labels in

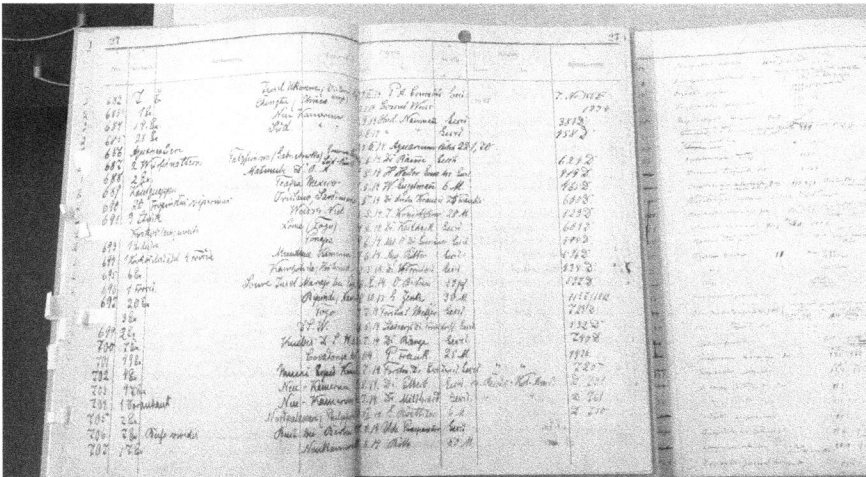

Figure 2.2 Book of accessions, Museum für Naturkunde Berlin. Photo: Author

the museum collections. They indicate a mutability of specimens and their order-ings that can take them across collections, classifications and purposes. A taxo-nomic revision might place an animal into the purview of a different curator or into another drawer while historical research might join crocodiles and birds as part of the same collection event. Like the GenBank record, then, pages from museum catalogues and registers can be read as bioinformational artefacts that point to the historical and material conditions embedded within current data-based logics. A noteworthy feature of the *Eingangsbuch* are the many transgres-sions of the handwritten entries as they spill out of the printed table. The "locality" column cannot contain the colonial place names which are noted diligently: "Chingtao" (Qingdao) in China, "Neu Kamerun" (New Cameroon) and "Süd Kamerun" (South Cameroon), Teapa in Mexico, Sardinia, "Weisser Nil" (White Nile), Lome in Togo, "Insel Marajo bei Para" in Brazil to "Nordpalawan" (North Palawan), Philippines. Neatness, the staying within the printed borders, gives way to an unbounded enthusiasm in recording the excesses of colonial extractions.

The *Eingangsbuch*, also in mirroring double-entry bookkeeping, is a testimony to the acquisitive spirit that has and continues to suffuse the collection of nature and data. The success of double-entry bookkeeping is commonly explained in terms of its "technical superiority" that "contributed to the historical emergence of a 'rational worldview'" (Carruthers and Espeland 1991: 33). In other words, the world became manageable because bookkeeping – its instruments and conven-tions – brought a specifically rationalized world into being. Stoler has made the argument that standardized records and more generally, bureaucracy, functioned as a central instrument for the colonial state *as a moralizing state* (Stoler 2009: 69 my italics). The orderly form provided by the printed layout provides a reflection of an order while at the same time providing the impetus for this order. In doing so, certain sensibilities such as the neat and orderly filling of forms and recording of entries become a virtue. In relation to the practice of 18th-century colonial nat-uralists and Defoe's Robinson Crusoe, the historian Anke te Heesen made the wonderful observation that both "turned to the compilation of tables to order their thoughts and keep their spirits up under challenging circumstances" (Heesen 2007: 238). The recording and ordering of specimens thus reflect the moral values and attachments of the people and institutions involved in it.

In the context of both databases and ledgers empty cells spell functional and moral trouble. Gaps, according to the sociologist James Aho (1985), are a sign for something amiss and cast doubt on whoever keeps the books. The most notable gap on the pages of the *Eingangsbuch* is the missing or in any case lacking data in the third column, *Bezeichnung* (description). This would be the place for the species name but instead the book provides only rudimentary identifications such as "Aquarientiere" (aquarium animals), "tadpoles", or "Krokodileingeweide" (crocodile innards). These omissions stand in stark con-trast to the detailed recordings of colonial place names. But they also reveal a rather weak attachment to science, that is, to recording the incoming animals using their scientific species names as scientific specimens. The book records transactions and documents stock rather than species.

Bioinformational artefact III: the paper

I have so far attended to two bioinformational artefacts, the GenBank record and the *Eingangsbuch*. More than 100 years stand between them, yet they return us to the history of natural history and to the materials (specimens, labels, registers) which underwrite dominant taxonomic classification and categories. Approaching them through data moments, momentarily released from the informational ecologies of database and museum collection, draws attention to their epistemic, institutional and social lives as products of and protagonists in bioinformation worlds. In this final data moment, I want to return to the barcode and to its intended purpose, that of allowing rapid and accurate identification of animal species.

The crocodile barcode that begins this chapter appeared in a paper entitled "Barcoding bushmeat: molecular identification of Central African and South American harvested vertebrate" by M.J. Eaton et al. This was published in 2010 in *Conservation Genetics*, a journal dedicated to genetic and evolutionary applications for the conservation of biodiversity, and is recorded in the GenBank record discussed above. In the paper the authors test the viability of barcoding to successfully identify and distinguish between processed animal parts (such as handbags and processed meat) of a set of "commonly hunted African and South American mammal and reptile species" (Eaton et al. 2010: 1389). At the same time, they are concerned with establishing barcoding as a "simple, low-cost and accurate" registration of species (2010: 1402). Their argument for the use of barcodes focuses on what they identify as a globalized bushmeat trade, which according to the authors affects species loss, human and agricultural health, international trade and legal frameworks such as CITES, the Convention on International Trade in Endangered Species of Wild Fauna and Flora. Conventional methods for species identification based on morphological traits are deemed too slow and unwieldy and are thus seen as impeding the control of illegal trading which requires definitive (legally certain) species designation. Indeed, the last ten years have seen a considerable increase in research advocating for the necessity and efficiency of DNA-based species identification in relation to "bushmeat".

In her wonderful study of DNA barcoding, the sociologist Claire Waterton (2010) closely examines the rhetoric employed by the barcoding community. Focusing on claims that barcoding will democratize taxonomy and offer standardized, swift and affordable access to species identification and species data, she draws attention to the critical work that the articulation of visions and promises does. Barcoding, she argues, is not just a "technique for ordering natural kinds" but in its promissory dimension entails "broader socio-political orderings" (Waterton 2010: 153). The term "bushmeat" already does its own type of ordering. A catchall for every kind of hunted meat including crickets, rodents and bats, it signals "a linguistic move that participates in the semiotics of denigration" by racializing people's subsistence (McGovern 2014). The target for species protection and its legal enforcement (as well as the burden of biodiversity conservation more generally) thus becomes firmly placed within the remit of states and societies that have been systematically impoverished by the very same operations and logics that filled museum collections with specimens of (endangered) species.

Another related question that Waterton's analysis opens up pertains to the nature of the socio-political problem barcoding is supposed to benefit. Biotechnologies do not just address problems but configure these problems in specific ways so that they become "doable" for the technology and its protagonists (Fujimura 1987). In the case of the Eaton paper and barcoding initiative more generally, biodiversity extinction becomes primarily articulated as an informational problem. Data is needed in order to identify species, which are the basic unit of biodiversity and thus compel measurement in order to quantify (the loss of) biodiversity. In turn, biodiversity become protectable by generating data on species, data that can be circulated across devices and domains. In this context, genetic data, whether in the form of barcode or whole genomes, becomes mobilized as legal evidence but it can also readily be commodified for bioprospecting. In literature expounding the virtues of genetic species identification for saving biodiversity, more data becomes the prudent imperative. And it is not coincidental that the datafication of nature has coincided with an understanding of environmental protection as "projects aimed at protecting natural resources (living and otherwise) for the sake of maximizing human utility" (Kim 2015: 8). The potential of data is continuously reproduced through ensuring conditions of re-combination, portability, mutability. Here, the perceived efficiency of barcoding also resides in its obfuscation of bodies, lives and histories. Colonial ruinations and racial capitalism do literally not compute as drivers of biodiversity loss. Framing biodiversity extinction as an informational problem thus distributes agency and risks in particular patterns that are also determined by access to the material infrastructures of barcodes as well as by the politics in and of species data.

Species: data creatures

In conclusion I want to return to the debate about the nature of sequence data and the demands to regard it as equivalent to its physical biological counterparts. The three interconnected data moments which I have described above – the database record, the *Eingangsbuch* and the Eaton paper – offer a small insight into the material counterparts and infrastructures that produce and maintain genetic barcodes. On one hand they show sequence data's non-linear, recursive paths that wind across times and places. Reconstructing these paths makes evident that barcoding, like cloning, is "an extension of an infrastructure that has been in the making for the better part of a century" (Landecker 2007: 231). On the other hand, the database record, *Eingangsbuch* and paper represent specific bioinformational artefacts that draw attention to the material location of data and the dependencies which stem from the particular forms that data takes. It is hard to define the boundaries around sequence data. The apparent simplicity of its code betrays a complex coming-together of times, sites, labours and infrastructures. Much has been written about the importance of metadata in furnishing data with the capacity to travel and make sense across communities. Yet, as the data moments show, bioinformational artefacts have always already been mobilized. Traces of their previous travels are to be found in the references, cross-references and citations as well as the odd marginalia that are found in catalogues and inventories. Instead of a straight referentiality between data and physical counterpart, their co-existence is characterized by a "cumulative relationality" that chains together data points from different sources (Nadim 2012). These sources – other databases, museum collections, international organizations, colonial expeditions, laboratories, sequencing technologies – represent critical sites in the location and accountability of data.

Ever since its introduction in the early 2000s, barcoding has attracted criticism from within biology and beyond (Abebe et al. 2011, Will et al. 2005). Some are worried that it might do away with classical taxonomy altogether, threating not just livelihoods but a longstanding expertise and its dedicated networks, institutions (such as museums) and cultures. Others are more concerned with the reductive nature of barcoding where one tiny genetic fragment is used as a universal marker for all species. These scientists advocate for an integrative approach to species identification that takes into consideration different markers, including but not limited to sequence data. What however unites taxonomists regardless of method is a desire to complete an inventory of life on Earth – this ambition is inherent to the discipline. The taxonomic ambition to name all species on Earth and its attachment to totality remains a central moral obligation within current data-based efforts to record and identify the world's biodiversity. And so, the spectre of "everything" structures the data-based efforts and the epistemic habits they compel.

According to Waterton, barcoding reconfigures species as "both a social object and a natural phenomenon" (Waterton 2010: 160) but given its history the logic of species has espoused socio-cultural hierarchies long before the advent of barcodes and DNA. Species is a concept, in fact, it is many concepts. These include, among others, typological species (based on shared set of fixed traits), biological species (based on reproductive isolation) and ecological species, which occupy a shared adaptive niche. Different communities work with different species concepts and so the effort to identify, name and classify life on Earth has always been a fractious one. Indeed, the history of science has produced many astounding accounts of the disputes over nomenclatures and categories and it has shown how these have also always been struggles over epistemic, moral and political authority (e.g. Schiebinger 2004, Schiebinger and Swan 2005). Yet, despite ongoing debate, the general consensus holds that the species category corresponds to something like *a natural kind* and that species are units of nature (e.g. DeSalle and Tattersall 2018, Maclaurin and Sterelny 2008). In other words, "species" is a social and historical construct and a powerful category of naturalized difference. Species are data creatures.[2] They are constructed and reconstructed in and with labels, registers, catalogues, database records and papers that draw together and materialize bodies, histories and information. Species are produced in data formations that involve museums and laboratories, colonial expeditions and algorithms. This is one reason for supporting the argument to treat digital sequence information in the same way as one would treat its physical biological counterpart: because they are not discrete entities, they are mutually imbricated through bioinformational artefacts.

Notes

1 See https://reptile-database.reptarium.cz (Last accessed: 18.05.21).
2 I borrow the term from Jennifer Gabrys's work on citizen sensing where she speaks of citizen-sensing projects as processes of *"creaturing data*, where actual environmental entities that come together are creations that materialize through distinct ways of perceiving and participating in environments" (2017: 13).

References

Abebe, E., T. Mekete & W. K. Thomas 2011. A critique of current methods in nematode taxonomy. *African Journal of Biotechnology* 10, 312–323.

Aho, J. A. 1985. Rhetoric and the invention of double entry bookkeeping. *Rhetorica: A Journal of the History of Rhetoric* 3, 21–43.

Carruthers, B. G. & W. N. Espeland 1991. Accounting for rationality: double-entry bookkeeping and the rhetoric of economic rationality. *American Journal of Sociology* 97, 31–69.

Costello, M. J., R. M. May & N. E. Stork 2013. Can we name Earth's species before they go extinct? *Science* 339, 413–416.

DeSalle, R. & I. Tattersall 2018. Evolutionary lessons. *Troublesome Science*, 1–24. Columbia University Press (available online: www.degruyter.com/document/doi/10.7312/desa18572-003/html, accessed 19 May 2021).

Eaton, M. J., G. L. Meyers, S.-O. Kolokotronis, *et al.*2010. Barcoding bushmeat: molecular identification of Central African and South American harvested vertebrates. *Conservation Genetics* 11, 1389–1404.

Foucault, M. 1970. *The order of things: an archaeology of the human sciences.* London: Tavistock.

Fujimura, J. H. 1987. Constructing 'do-able' problems in cancer research: articulating alignment. *Social Studies of Science* 17, 257–293.

Gabrys, J. 2017. The becoming environmental of computation from citizen sensing to planetary computerization. *Italian Journal of Science & Technology Studies* 8, 5–21.

Hebert, P. D. N., A. Cywinska, S. L. Ball & J. R. de Waard 2003. Biological identifications through DNA barcodes. *Proceedings of the Royal Society B: Biological Sciences* 270, 313–321.

Hebert, P. D. N. & T. R. Gregory 2005. The promise of DNA barcoding for taxonomy. *Systematic Biology* 54, 852–859.

Heesen, A. te 2007. Accounting for the natural world: double-entry bookkeeping in the field. In *Colonial botany: science, commerce, and politics in the early modern world* (eds) L. Schiebinger & C. Swan, 237–251. Philadelphia, PA: University of Pennsylvania Press.

Jones, J. E. 2011. *In search of brightest Africa: reimagining the dark continent in American culture, 1884–1936.* Athens, GA: University of Georgia Press.

Kim, C. J. 2015. *Dangerous crossings: race, species, and nature in a multicultural age.* Cambridge: Cambridge University Press (available online: www.cambridge.org/core/books/dangerous-crossings/F9B0757D753463291E59D99065C5C193, accessed 20 May 2021).

Knorr-Cetina, K. 1999. *Epistemic cultures: how the sciences make knowledge.* Cambridge, MA: Harvard University Press.

Landecker, H. 2007. *Culturing life: how cells became technologies.* Cambridge, MA: Harvard University Press.

Leonelli, S. 2010. Packaging small facts for re-use: databases in model organism biology. In *How well do facts travel?: The dissemination of reliable knowledge* (eds) P. Howlett & M. S. Morgan, 325–348. Cambridge: Cambridge University Press.

Lewin, H. A., G. E. Robinson, W. J. Kress, *et al.*2018. Earth BioGenome project: sequencing life for the future of life. *Proceedings of the National Academy of Sciences* 115, 4325–4333.

Maclaurin, J. & K. Sterelny 2008. *What is biodiversity?*Chicago: University of Chicago Press.

McGovern, M. 2014. Bushmeat and the politics of disgust. *Society for Cultural Anthropology* (available online: https://culanth.org/fieldsights/bushmeat-and-the-poli tics-of-disgust, accessed 20 May 2021).

Nadim, T. 2012. *Inside the sequence universe: the amazing life of data and the people who look after them*. PhD Thesis, Goldsmiths, University of London, London, UK.

Nadim, T. 2016. Data labours: how the sequence databases genbank and embl-bank make data. *Science as Culture* 25, 496–519.

Nadim, T. 2021. The datafication of nature: data formations and new scales in natural history. *Journal of the Royal Anthropological Institute* 27, 62–75.

Ratnasingham, S. & P. D. N. Hebert 2007. Bold: the barcode of life data system (www. barcodinglife.org). *Molecular Ecology Notes* 7, 355–364.

Richards, T. 1993. *The imperial archive: knowledge and the fantasy of empire*. London: Verso.

Savolainen, V., R. S. Cowan, A. P. Vogler, G. K. Roderick & R. Lane 2005. Towards writing the encyclopaedia of life: an introduction to DNA barcoding. *Philosophical Transactions of the Royal Society B: Biological Sciences* 360, 1805.

Schiebinger, L. L. 2004. *Plants and empire: colonial bioprospecting in the Atlantic world*. Cambridge, MA: Harvard University Press.

Schiebinger, L. L. & C. Swan (eds) 2005. Colonial botany: science, commerce, and politics in the early modern world. Philadelphia, PA: University of Pennsylvania Press.

Stoler, A. L. 2009. *Along the archival grain: epistemic anxieties and colonial common sense*. Princeton, NJ: Princeton University Press.

Vismann, C. 2016. Out of file, out of mind. In *New media, old media: a history and theory reader* (eds) W. H. K. Chun, A. W. Fisher & T. Keenan, 97–104 (Second edition). New York, NY: Routledge.

Waterton, C. 2010. Barcoding nature: strategic naturalization as innovatory practice in the genomic ordering of things. *The Sociological Review* 58, 152–171.

Will, K. W., B. D. Mishler & Q. D. Wheeler 2005. The perils of DNA barcoding and the need for integrative taxonomy. *Systematic Biology* 54, 844–851.

3 Capturing genomes

The friction and flow of bioinformation at the Smithsonian

Adrian Van Allen

Introduction

On a cold morning in March 2015, I stood in the Smithsonian Institution's Biorepository holding an insulated box as my colleague swiped his badge to open the door. Within the box I held were stacks of smaller boxes holding tissue tubes, samples from species around the world that were being collected and collated into a genomic archive of life, now being sorted and placed in liquid nitrogen filled tanks to preserve them in perpetuity. In this vast off-site facility in rural Virginia, the frozen spaces were clustered together—the Biorepository next to freezers full of cancer-riddled kittens, part of a long-term study for the National Institute of Health, adjacent to the film vaults for the National Museum of American History, stacked full of reels of flammable silver nitrate film. These various frozen materials were being kept cold for perpetuity, frozen for the future while at the brink of dissolving into something else, decay kept at bay.

Setting down our tissue sample box at a workstation, we opened it and removed the two tubes that were to play a vital role in determining the data standard for what would be included or excluded in this ever-growing genomic archive—slivers of muscle tissue from a White perch (*Morone americanus*) and a Chesapeake blue crab (*Callinectes sapidus*) in two slim plastic cryotubes, a barcode label affixed to their sides (Figure 3.1). Logging into a computer at the workstation we scanned the barcodes of our fish and crab, noting what their location would be within the rows of massive stainless-steel tanks behind us. Together, the tanks had a capacity to hold up to 5 million tissue tubes suspended above a pool of liquid nitrogen to keep the samples frozen at -190°C. Donning lab coats, face shields and large blue padded gloves we climbed up the steps on the side of the tank, opening the lid to withdraw a rack that held the frozen tubes in place (Figure 3.2). Brushing aside the frost on the rack to read the numbers, we slide our tray of fish and crab into its allotted space, resealing the tank lid as quickly as possible to maintain the internal temperature and keep the tissue tubes in their state of suspended animation (Figure 3.3).

I begin at the end for this story, with the placement of these two tissue tubes into their final resting place within the Smithsonian's Biorepository. This

DOI: 10.4324/9780367810030-3

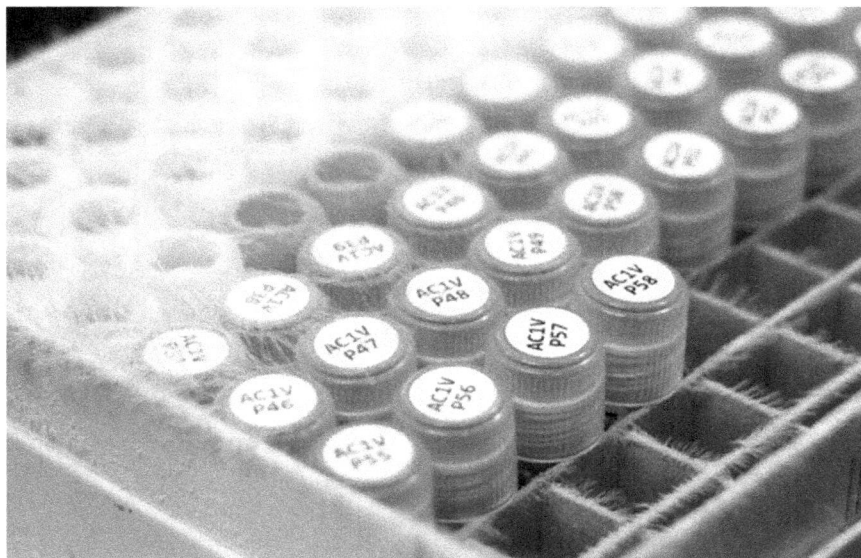

Figure 3.1 Frozen tissues floating in a tray of liquid nitrogen (Smithsonian Biorepository, March 2015)

Figure 3.2 Empty racks inside a liquid nitrogen tank ready to fill with tubes (Smithsonian Biorepository, March 2015)

Figure 3.3 Rows of stainless-steel tanks full of frozen genetic samples (Smithsonian Biorepository, March 2015)

moment was but one link in a long chain of events in the life history of two animals that became specimens—specimens that would mark the genome-quality standard for amassing one of the largest tissue collections in the world, currently being built to put "all life on ice" for uncertain ecological futures. This chapter examines the negotiations and practices of transforming life into data at the Smithsonian National Museum of Natural History in Washington D.C., focused on the Global Genome Initiative (GGI). The GGI's self-described project is to create a "genome-quality" frozen archive of half of all taxonomic families of all life for an uncertain ecological future within the next six years, collaborating with a global coalition of over 100 other museums, herbariums, biorepositories and DNA banks. What constitutes a "genome-quality" tissue sample is at the center of this project, entangled with the layers of infrastructure and politics of integrating genomics into the museum context.

The site of this ethnographic inquiry is within these two tubes, as I seek to unravel how much meaning is condensed into such a small object, these simple 2ml vials of plastic that can contain multiple imagined futures bound up within the tissue samples they contain. The material practice of creating these two unassuming tissue tubes is one of "rendering flesh into data" (Radin 2012, 310), following the biologies of these creatures as they are fractioned into new kinds of museum objects that are capable of both carrying and reproducing certain kinds of information, including negotiating certain kinds of (genomic) futures. Attending to how these tissue tubes came into being, I examine how the material properties of their original animals – a fish and a crab – then become a standard for measuring the relative quality of other tissue tubes taken from other animals collected across the globe in a mass salvation effort. This chapter will focus on these processes, or series of transformations from animal to specimen, from specimen to data, and finally into a data standard for future specimens. In doing so, I want to suggest that engaging first-hand with the material practices of museum genomics is essential – that is, the site of this ethnographic research combines views gleaned from the work bench and in the lab meetings where genomic tissue standards are negotiated. The standardization processes of making a variety of specimens fit into a single data schema are based not only on making data standards uniform across disciplines, but are also, I argue, profoundly shaped by the material practices from which those data standards arise.

Based on three ethnographic episodes at the Smithsonian Institution in Washington, D.C., I argue that the material practices preceding the movements of data fundamentally shape and inform the paths which a specimen can take. Save the entire fish and you have a traditional voucher, but preserve it in formalin before you take a tissue sample, and you won't be able to get DNA from it. In short, biodiversity conservation through genomic collecting orients museum sociologies, biologies and ecologies—continually engaging them in ongoing processes of remaking, re-inscribing, or perhaps removing (if indeed they ever existed) the boundaries between nature and culture. Engaging the material practices of how archives are made, I examine how these negotiations for knowledge standards are made manifest through the biomaterials of the specimens themselves, as they are unraveled from wholes and into parts. I suggest that it is within this fragile network of negotiations that different forms of power are rendered visible. Further, it is the importance of the biomaterials themselves in making the data standards that comes to the fore in these encounters—what "behaves" or "misbehaves," that is, what to do with specimens and their data that don't fit into pre-existing categories, where life continues to overfill the preset frames of reference. It is these moments of alignment and misalignment, of multiply layered frames of reference that bind these specimens and their samples together, but not always in unison, where we can see the friction and flow of bioinformation in and through the genomic museum.

I began in the Biorepository, sorting the tissue tubes into their identified slots, each tube serving as the voucher for the genetic information that was extracted from them—the physical anchors in the chain of information binding a physical specimen in a museum cabinet, drawer, or liquid nitrogen tank to the abstracted genetic data that lives on servers and circulates beyond the specimen it originated from. These frozen tubes in their numbered rack were the ending point of the specimens after they had been collected, processed, sectioned into parts and pieces, and portioned into various tissues tubes. One of these tubes went to the Biorepository, another to the Laboratories of Analytical Biology at the Smithsonian National Museum of Natural History, to be consumed as the DNA was extracted and measured. This is my second encounter at the lab bench extracting DNA and assessing its viability for being used as standard for "capturing genomes" by global partners at other museums. Finally, my third encounter is in a lab meeting to discuss the results, and ultimately defining the Global Genome Initiative's genome-quality data standard.

Bringing into question the contemporary standardization of biodiversity to make it knowable, computable and sharable, I examine the Smithsonian's Global Genome Initiative (GGI) creation of a "genome-quality" tissue standard. Through exploring the reassessment of natural history collections as biological libraries of "life's code" and as the setting for the standardization of biodiversity data, the drive to create an entire corpus of human knowledge of life through, as one of the Smithsonian scientists phrased it, "putting all life on ice." This urge to archive all life connects issues of care for biodiversity, as well as its potential loss—as taken from one specific and culturally situated perspective, a perspective that fundamentally orients collection strategies. Within this context, I consider the way that the biodiversity biobank-as-archive holds together particular ideas about future orderings of nature and culture, and facilitates new collaborations around the unraveling biological objects in its care.

To begin to answer these questions I turn to longer histories of specimens-as-data within the museum. Tracing back to seventeenth-century cabinets of curiosity, I examine how every age has been one of "data deluge," faced with the challenge of incorporating new categories of life into existing classification systems. Following these histories of standardizing biodiversity in museum work, I connect ongoing debates on creating a "genome-quality" tissue standard to the friction of and flow of biodiversity as specimens move into the Smithsonian National Museum of Natural History collections—points in a trajectory towards organizing biological life into an abstracted open-source data project.

The material practices of constructing genomic archives

Scholarship on the creation of archives has looked to the forms of power created in the hierarchies of knowledge (Bowker and Star 1999, Daston 2004, Durkheim and Mauss 1963, Ellen 1993, Lampland and Star 2009, Needham 1979). Instead I focus on the material practices that bring these forms of data into being, moving back along the trajectory of the specimen to the inflection point when it is transformed by the labor and thought of specimen preparators and lab technicians, moving from meat to meaning, from matter to immateriality. Further, I want to place these material practices of museum genomics in their historical context, where the generation of new types of objects and their associated data has always been a problem for collections, dating back to early cabinets of curiosities (Strasser 2012).

Previously, I've examined the shifting value of natural history collections and their articulation as untapped resources (Van Allen 2018, 2019), looking at the conceptual framing of collections as "biological libraries," that is, as sites to extract data relevant for fields such as agriculture, national security, disease control and as storehouses of knowledge for biodiversity conservation. One aspect of conservation is preservation, which I argue can be understood as a return to eighteenth-century encyclopedic collecting now utilizing contemporary genomic tools. Frictions between these two orientations—to conserve for future use or to preserve in perpetuity—are brought into focus through the daily practices of crafting data standards for making "genome-quality" tissue collections, or what the scientists at the Smithsonian called "capturing genomes" (Droege et al. 2016, Mulcahy et al. 2016).

As making collective meaning through classification continues across cultures (Bleichmar and Mancall 2011, Levi-Strauss 1966), it also changes in materials, forms, and the details of how a classification system is "anchored" over time. This includes, in my particular context of the contemporary natural history museum, the "anchors" of physical specimens tied to the genetic data derived from them. For taxonomy, the value of biological specimens and the natural order(s) they are taken to represent are upheld by meticulously observed chains of connections vitally linking the specimen to information, to nature, and back again. These connections are built from data—data that is derived from the biomaterials of the specimens themselves, from what is preserved versus what is discarded. Each discipline values and discards different pieces, as seen in one of the parasitologists I worked with defining her "field site" as the intestines of birds, a site where she could extract worms with the tips of two needles, utilizing a part of the specimen that would have been thrown away by the ornithologists at her museum.

While the classification of nature has been acknowledged as a central human activity (Foucault 1966, Levi-Strauss 1966), and one shaped by hopes for its protection on the one hand and exploitation on the other (Hayden 2003, Lowe 2006, Tsing 2005), taxonomy as an endeavor continues to be perceived as a rarified and esoteric set of knowledge-generating practices. However, one consequence of the signing by over 150 nations of the Convention on Biological Diversity (CBD) in 1992 was an unprecedented focus of global attention upon the significance of taxonomic knowledge as an underpinning prerequisite for the protection of an ever-dwindling global biodiversity. The United States, it should be noted, was not one of the signatories of the CBD. However, I was assured by Smithsonian staff that following the Convention on Biological Diversity, the Nagoya Protocol's best practices for collecting and curation are "not only the legal thing to do, even if we aren't a signatory, it's the right thing to do." Biodiversity, then, is shifting from being defined simply as a network of all living things to being defined as a (continually emerging) network of interests negotiated between nations, institutions, and individuals. Museum collectors and their ever-growing genomic archives are but one stakeholder in this web.

Extending the collections: capturing genomes

January 2015. Another snowy morning, now a year after the Smithsonian's Global Genome Initiative was launched as a new project to collect and centralize all the tissue samples from across the various museum departments. I sat in the long white conference room in the Laboratories of Analytical Biology

(LAB) at the Smithsonian National Museum of Natural History. Six of us were at the table, all heads turned towards the monitor at the end of the room that displayed two DNA gels (Figure 3.4 and Figure 3.5). The DNA gel images are

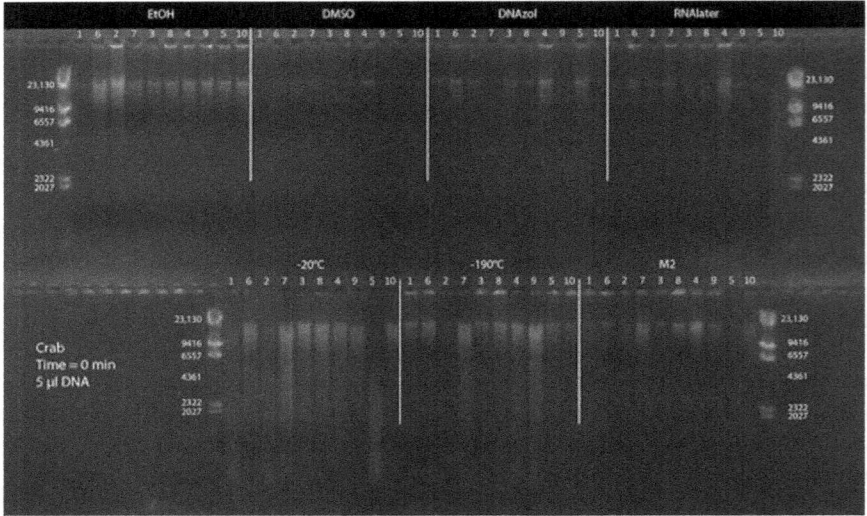

Figure 3.4 Crab DNA gel image (Smithsonian Laboratories of Analytical Biology, January 2015)

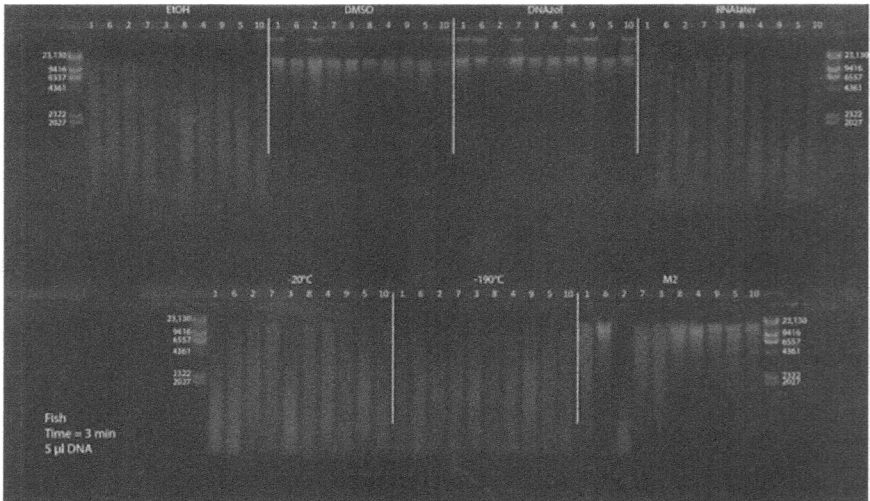

Figure 3.5 Fish DNA gel image (Smithsonian Laboratories of Analytical Biology, January 2015)

two dark gray rectangles, filled with pale smudges in ordered rows, each row marking the progress of extracted DNA through the thick gel under a low current. The brighter the small square smudges, the higher the molecular weight of your DNA, and therefore the higher the quality of your samples. It is, I come to learn, a matter of contrast.

The goal of this meeting was to finalize a scientific paper titled "Capturing Genomes," which when published will serve as a protocol for how to measure the quality of the DNA in one's collections—be they collections in the NMNH Biorepository or one of the (at last count) 34 collaborating institutions world-wide that are part of the Global Genome Biodiversity Network (GGBN). The Capturing Genomes paper had been through many iterations, and was a core piece of the puzzle for the GGI to move forward with its goals—a standard was needed to determine the quality of the collections being made in the GGI's name, and across the other collecting activities of the GGBN. Since the GGI's stated purpose is "Preserving and Understanding the Genomic Biodiversity of Life on Earth," it was key that what was being collected and preserved was in fact "genome quality." Precisely what determined "genome quality" was under scrutiny as well, and went through a number of iterations, but the "working answer," as one curator put it, was to "settle on a DNA fragment length of 9kb (kilobase pairs), for the time being," since "things change so quickly."

The meeting today was to go over the remaining issues about the paper. These included looking at the gel images that had been run by the GGI lab tech of a fish and a crab, the two test cases for the paper, who we have already encountered at the beginning of this story, as they were sorted into the frozen time of the Biorepository. And then there was the issue of vouchers—the specimens that would go into the collections as the reference for the tissues of the fish and the crab—ideally a whole identifiable organism that would become part of the Departments of Ichthyology for the fish and the Department of Invertebrate Zoology for the crab. For now, we collectively stared at the screen, looking at the smudges of black and gray on the DNA gel. The gel images were not as hoped, and there were many concerned expressions and the occasional frustrated sigh. The paper was overdue, but the results from assessing the amount of DNA in the samples didn't seem viable.

As we examined the gels on the screen in front of us, the fish and the crab were pale smudges compared to the "ladders," or controls that ran along on either side—the reference so one could gauge that the test worked. Beyond the paleness of the smudges, and far more problematically, they were inconsistently pale. The trouble with the fish and the crab was their variability—the same sample "run" on two gels, mixed and cast at the same time, returned wildly different results. The general consensus (after a careful recounting of the process to make sure an error hadn't been made) was that the results were unusable.

"We're making a standard for everyone in the GGBN [Global Genome Biodiversity Network] to use, it has to be rock solid. And this isn't it," stated an Invertebrate Zoology curator. The protocol being put forth in the paper was going to be integrated into the collecting practices of the full network of collaborators, over 100 global collaborating institutions each with varying degrees of funding and access to advanced tools. Questionable data were not acceptable. Debate ensued, going over the details of the numbers, re-doing the calculations, asking what the best way was to standardize against the opacity of the gel, how to equalize the contrast in the image, visiting the possibility of re-doing it all from scratch. Time, funds, labor and accessibility of specimens were all factors against re-doing all of it from the beginning. Not to mention, as one curator pointed out, the need to get the protocol tested and out into the GGBN community so they could begin adding to the genomic archive as quickly as possible in the face of mass biodiversity loss. Conversation lulled into silence.

"What's the standard deviation for the TapeStation?" the Invertebrate Zoology curator asked into the silence. Heads popped up from behind sheets of paper and laptops. More debate ensued, but with a renewed pitch of excitement. This might be a way through. To ensure that genomic DNA is of the quality required, the group agreed, each sample needed to be screened to determine its suitability before committing the time, money and resources to preserving it in a biorepository, and taking up space in the collections for its voucher. Assessing genomic DNA is usually performed by agarose slab gel. However, this is a slow, labor-intensive, and manual process that can take several hours. In contrast the largely automated workflow of the TapeStation uses a credit-card-sized device made of three separate polymer layers that separate biomolecules through a gel matrix in separated channels. Genomic DNA is mixed with buffer, and then placed in the TapeStation for automated analysis. A "plug-and-play way to get your DNA's molecular weight" as a lab tech interjected into the conversation.

In this narrative the traditional agarose slab gel is considered outmoded, slow, and ultimately somewhat unreliable given its handmade quality. However, the very cheapness of the handmade slab gel is precisely what makes it appealing for a global collaboration such as the Global Genome Biodiversity Network (GGBN), stretching across institutions in many third world countries located in biodiversity hotspots that were desirable for the genomic archive, but these institutions had varying amounts of labor, funding and equipment available. Microwaves to heat up buffer and melt agarose gel, UV lights, and electricity to run a traditional agarose gel were usually accessible to those with even limited means. TapeStations were not. A quick scramble as heads bent back to their laptops, clicking away as the variability rates for the TapeStation were looked up from previous projects. "The variability of the results of the fish and crab aren't really different than the TapeStation," another lab tech offered up. "I think we're OK." The mood in the room swung in a sea change, it was almost jubilant. The details of where and how to word the protocol were nailed down; the process of collating edits and comments into one document organized. Another lull in the discussion, and someone asked, "What about the vouchers?" More silence as various gazes locked across the room.

The original specimens had been collected off the coast of Panama, at the Smithsonian Tropical Research Institute (STRI).[1] This was an efficient way to collect specimens, I was told, with a minimum of permits and paperwork as the research station and collections made within its boundaries are considered a priori part of the Smithsonian. "Were they collected before October 14, 2014?" a GGI staff member asked one of the lab techs. He checked the paperwork and nods. "Good, before Nagoya," a reference to the Nagoya Protocol, a piece of international legislature that came into effect during my fieldwork at the museum, a refinement of the Convention on Biological Diversity from 1993. In defining the biowealth sovereign nations and their control over their flora and fauna, it also fundamentally changed the political topography of moving specimens across international borders.

According to scientists I interviewed across the Smithsonian, specimens were getting much more difficult to get, to move, and to keep. The crab and the fish, however, predated the entanglement. Not much of them was left, however. Apparently, the fish were collected as "little fish fillets" into a jar of ethanol, according to one of the lab techs, and the only remains of the crab was the left claw. "We need a phylogenetically valid voucher somewhere," one of the GGI staff says, "at least two tubes in the biorepository, if nothing else. If there's a claw in a jar in IZ [Invertebrate Zoology], fine, but we need to have these live on GGBN before [the paper] gets published." Heads nod. "Voucherless tissues?" someone asks. The carefully connected links in the chain between the original whole fish and crab, their tissue samples, and their extracted and abstracted DNA data were starting to come apart. A pause in the conversation as various minds imagined that chain and how to rebuild it. "The vouchers are the tissues," one of the curators replies, and pauses before adding "for now." The possibility of getting a substitute voucher for the fish tissue is discussed briefly—what's referred to as an allotype, one of the same species (usually verified through DNA barcoding, as well as being a visual, morphological match), but not the specific individual that the tissue sample came from.

Population sampling is a common practice in various Divisions and Departments across the Smithsonian, as it is in many other natural history museums. For example, Invertebrate Zoology collectors often sample various marine creatures in a colony, such as a coral polyp colony living in a reef, where many are samples, but one is picked to serve as a voucher. Similarly, collectors for the Botany Department take samples from fields of the same plant and then press one full individual as their voucher specimen. The relationship of one-to-many in the collections made in these disciplines runs counter to the one-to-one relationship in other disciplines, such as Vertebrate Zoology. For the Division of Birds, a part of Vertebrate Zoology, where many of the first genomic collections were made 20 years ago, the ideal relationship remains as a whole, stuffed bird skin in a drawer, with tissue tubes sampled from the heart, muscle and liver stored in the Biorepository, extracted and replicated DNA from those tissues used in the biolab and then sent to join their tissues in the Biorepository, and all the associated genomic data filed into several online databases. Various extraneous pieces can unravel from these specimens—the parasites in the intestines are of interest to the invertebrate zoologists for their National Parasite Collection—but the essential links in the chain of whole specimen to tissue to DNA to data remain unbroken. The fish's replacement voucher specimen would just be a bit out of sequence—tissue first and allovoucher second (at some point in the future). Someone suggested putting the jar of "little fish fillets" preserved in ethanol in the Division of Fishes as the voucher. One of the curators grimaces, shaking his head, this solution seems to offend his sensibility of what belongs in a collection. One of the lab techs agrees. "It just looks like meat," he says with a dismissive note, "why would it go into the Ichthyology collections? Just put it in a tube in the Biorepository—that's where it belongs. Meat in a tube" (Figure 3.6). That seems to resolve the issue. Notebooks and laptops are closed, and we shuffle out, turning off the monitor with its fading glow of DNA gels. They, too, have been sorted into their appropriate category of unexpected results, which though at first considered too variable, were then rendered into results that were no more variable than other practices already in place.

Standards were kept, boundaries of acceptability negotiated and maintained, vouchers found, or a future slot positioned for them. The fish and the crab were misbehaving, or more precisely the "read" on the molecular weight of their DNA was misbehaving (and certainly their vouchers were misbehaving). However, they were not misbehaving substantially more than the TapeStation or other vouchering methods. Therefore, it was an acceptable amount of misbehavior and could be

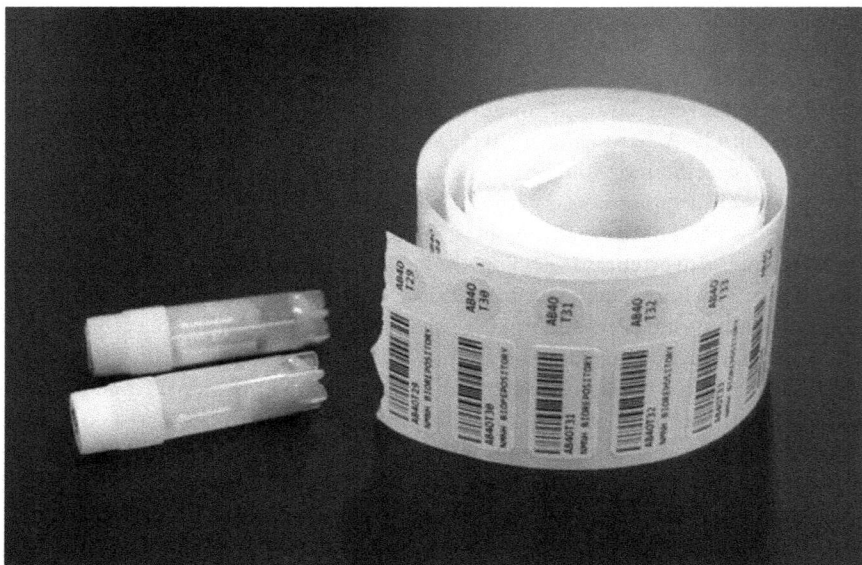

Figure 3.6 Tissue tubes ready to be labelled with Biorepository barcode stickers (Smithsonian Biorepository, March 2015)

accommodated. Despite their resistance, standards were being crafted to accommodate their idiosyncrasies and make them do the work necessary—that is, provide a method for "capturing genomes" for the Global Genome Initiative and its partners.

From meat to meaning: (frozen) cabinets of curiosity

We now return to the Biorepository, carrying our precious tubes of fish and crab, ready to be sorted into the trays afloat above liquid nitrogen. Through following their journey from meat to meaning we can see them transform in front of us. A tiny frost-covered cryovial becomes link in chain binding together its data on the servers, to the data set associated with the "Capturing Genomes" paper, to the small pieces deposited in the Departments of Invertebrate Zoology (but not Ichthyology). A crab claw, it seems, is worthy of space on a museum shelf, but a tiny scrap of fish flesh in a jar is not. These kinds of sortings of matter and meaning have a long history, of new kinds of objects emerging and being categorized.

Data within the museum has always been a central concern, and I began to think about both natural history and the larger histories of nature in the context of genomic collecting, its data and expanding networks of tissue tubes. What began as studies of local flora and fauna in sixteenth-century Europe was soon complicated and broadened by objects, specimens, and even living humans brought back to Europe from explorations to the New World and traders returning from expeditions (Daston and Park 1998, Findlen 2002, Greenblatt 1992, Impey and MacGregor 1985, Olmi et al. 2001, Pomian 1990). Collecting provided a means not only to assemble the newly discovered, but to make sense of it and exert control over its natural resources—a cabinet of curiosity built in 1648 demonstrates a complete cosmos in miniature, combining coral and shells from expeditions with a clock, biblical scenes carved into semi-precious stones from around the world, and a hidden pharmacy with jars carved from African ivory. The world remade from the material evidence of colonial domination, imprinting Western scales of time, Christianity and medicine on those representative fragments of coral, gems, and ivory. These collections within these early modern cabinets of curiosity formed the basis for many contemporary museums, and in perhaps some deep-seated and unexpected ways, they have also formed the basis for how the miraculous and the mundane are standardized to fit within a particular ontological schema. The freezer in the Laboratory of Analytical Biology, where our fish and crab tissues were stored while their DNA was being sampled and extracted, features an array of global diversity condensed into a new form of curiosity cabinet, one full of spiders from Costa Rica, fish from Timor, mammals from Brazil, snakes and lizards from Myanmar.

In addition, newly emerging technologies in the seventeenth century facilitated the preservation, circulation, and documentation of collections in new ways—such as specimens preserved in alcohol, and an outpouring of books that cataloged, categorized, inventoried and illustrated the collections in printed catalogs that could circulate far further than the stuffed, pinned, and pickled specimens they represented (Findlen 1994, Zorach et al. 2005). Massive collections spanning fields of knowledge and new areas of the world were amassed, organized, displayed, and circulated. Laying a claim to the value(s) of biological specimens thus raises a whole range of questions concerning what a specimen can ultimately stand for and forces us to imagine what it might mean to scale-up from a specimen in a mahogany drawer in the sixteenth century or in a liquid nitrogen tank in the twenty-first century to an appreciation of (potentially multiply constituted) "life itself" (Rose 2009).

Objects are made meaningful according to how they are placed within relations of significance. These relationships, in turn, depend on who is determining what counts as significant. Objects are therefore likely to be spoken, rather than to speak (Haraway 1997). This is not to say that all meanings are of equal value, of the same power, or of the same validity. We see this in the shifting interpretation of a crab broken into its biomaterials, as it is unraveled into different parts and pieces with each transformed to carry different weights of meaning with them—be it a claw in a jar, a tissue sample frozen in a tube, extracted DNA expanded across a gel, or its DNA barcode uploaded to a database.

Conclusion: the friction and flow of bioinformation

The move towards a stable (taxonomic) ontology of biodata and the corresponding claims towards the "data deluge" of contemporary science obscures a much longer history of biological collections as data sources. In other words, current museum genomics does not take into consideration the long history of museum collections as sites for data extraction—museums have always been "data banks," and each era is one of "data deluge" (Strasser 2012). As the life sciences increasingly become the biggest of Big Data projects (Leonelli 2013, 2014, Page et al. 2015), it is crucial to contextualize them in a longer genealogy of museum collecting.

The entangled histories of natural history museums and bioscience research can be followed through the divestment and reintegration of lab science in the museum in the early twentieth century to the present, a timeline that also corresponds to the emergence of anthropology as a discipline. However, in this chapter I have looked further back to the origins of the natural history museum, examining the shifting assemblages of specimen-as-data and the global networks of living (and formerly living) things in the first natural history collections in early modern European cabinets of curiosity. In contrasting these with contemporary genomic collecting, I have begun to think through the continuities and ruptures in the material practices of nature-making in museums.

Museum genomics is merely the most recent iteration of viewing the natural world as a data set to be collected and analyzed. Framing natural science collections as databanks reconfigures the collections as valuable to expanded audiences, transforming them into resources—and potential solutions—for contemporary crises, both social and biological. These include more obvious projects such as biodiversity conservation in the face of mass species extinction, as well as less obvious projects such as agriculture negotiating the influx of invasive species, national security dealing with invasive-species-as-potential-bioweapons (Dudley and Woodford 2002), disease control by charting contagion vectors from historic specimens (Suarez and Tsutsui 2004), and even the improbable de-extinctioning of species such as passenger pigeons (Revive and Restore 2013) and mammoths (Church and Regis 2012, Poinar et al. 2006, Shapiro 2015, Zimov 2005). However, it is also worth considering what kinds of labor and interests are involved in reconfiguring collections as data resources, and as we have seen, the handcrafted standards that shape these collections.

Different forms of life, including biodiversity, have been increasingly defined by molecular biology (Bowker 2000, Keller 2009, Sunder Rajan 2006)—where the metaphor of DNA as a code (Kay 2000) or a text (Ridley 2000) to be written or rewritten replaces the sticky materiality of the thing itself (Tsing 2005). Attending to the gap between "genetic 'information' and biological meaning" (Keller 2009: 7–8) is one productive way to think through how specimens become data. The details of an organism's genome, its parts and interrelated functions increasingly define what it means to be "alive" in the contemporary moment, as determined by human biomedical and biodiversity genomics and then dispersed and "naturalized" into larger cultural domains (Genome 10K Project 2016, Kowal, Radin and Reardon 2013, Parry 2004).

Species, in this conceptual framework, become their genomes in one sense—the protein sequences extracted from frozen samples, "read" and sorted into the "book of life" (Canguilhem 2008, Kay 2000, Ridley 2000), ready to be read again or rewritten as needed with emerging technologies such as genome engineering (CRISPR/Cas9 Guide 2016). From this standpoint "capturing genomes" and collecting "all life on ice" become plausible endeavors, based on a view that life is reducible to a 2ml cryovial, the tissue inside it, and the genomic data that can be extracted from it. The diversity of biodiversity is in the process of being transformed into stable, standardized categories to enable its collection, preservation, analysis and use within an existing ethos of "natural order"—an ethos that privileges the rarity of the species, the high molecular weight of the sample, the analytical chain of permits and vouchers, and the accessibility and visibility of the genomic data to institutional infrastructures or partner networks (such as the Global Genome Biodiversity Network)—all pointed towards preservation for an uncertain future.

To return to the shifting value of biomaterial as it is transformed into bioinformation, I want to highlight the creation of a "genome-quality" tissue standard as a way of thinking through things (Henare, Holbraad, and Wastell 2007; Ingold 2010). The ways these standards and the objects they organize come-into-being (Gosden and Marshall 1999, Moutu 2006) shifts the focus to new conceptual linkages between the lives—and afterlives—of organisms, both human and nonhuman, as they circulate. My ethnographic engagement with biodiversity collecting in the museum and the complicated webs of interaction that define it—socially, biologically and ecologically—relate to Helmreich's work on collecting the scientists who collect aquatic microbial life (2009). His concern with the microscopic, molecular, and genomic explorations of the open ocean and deep-sea point to an expansion of the concept of bios, of life in the alien ocean as other life. While his focus centers on engaging the scientists on their own particular spaces and their own particular terms—an orientation to ethnographic practice I find compelling—his collection of scientists offers up a somewhat homogenous narrative of biodiversity salvation. In contrast, the creators and collectors of museum genomics in my own research offered a variety of narratives based on their own disciplines' distinct histories of collecting and preserving, which they struggled to fit into a standardized schema.

From this perspective, the scientists I worked with are in the process of transitioning from being stewards of life's diversity in distinct disciplinary ways to becoming the conduits for increasingly standardized versions of life as they integrate genomic collecting practices, such as the protocol detailed for capturing genomes of a specifically high molecular weight. Standards make data accessible, but they also draw invisible lines between what is kept and what is discarded, naturalizing the remaining data, practices, specimens and interests and obscuring the labor required to make and maintain them. In other words, biodiversity conservation through genomic collecting orients museum sociologies, biologies and ecologies—continually engaging them in ongoing processes of remaking, re-inscribing or removing the boundaries of nature and culture. In turn, these nature-culture assemblages have the potential to expand the multiple possibilities for thinking about human and nonhuman relationships as we move into ever-uncertain futures.

Note

1 The Smithsonian Tropical Research Institute (STRI) is a bureau of the Smithsonian located on Barro Colorado Island in the Panama Canal Zone "dedicated to understanding biological diversity" (STRI 2016). Begun in 1923 as a small field station, the current institute's research activities extend across the tropics. These include STRI's Center for Tropical Forest Science that uses labeled forest plots to monitor tree demography in fourteen countries located in Africa, Asia and the Americas, and STRI marine scientists conducting a global survey of levels of genetic isolation in coral reef organisms as well as providing fish and crab specimens for the GGI (STRI 2016).

References

Bleichmar, Daniela, and Peter C. Mancall. 2011. *Collecting Across Cultures: Material Exchanges in the Early Modern Atlantic World*. Philadelphia: University of Pennsylvania Press.

Bowker, Geoffrey. 2000. Biodiversity Datadiversity. *Social Studies of Science* 30 (5): 643–683.

Bowker, Geoffrey, and Susan Leigh Star. 1999. *Sorting Things Out: Classification and Its Consequences*. Cambridge, MA: MIT University Press.

Canguilhem, Georges. 2008. *Knowledge of Life*. Edited by Paola Marrati and Todd Meyers. New York: Fordham University Press.

Church, George M., and Ed Regis. 2012. *Regenesis: How Synthetic Biology Will Reinvent Nature and Ourselves*. New York: Basic Books.

CRISPR/Cas9 Guide. 2016. CRISPR/Cas9 Guide. 2016. www.addgene.org/CRISPR/guide.

Daston, Lorraine. 2004. Type Specimens and Scientific Memory. *Critical Inquiry* 31 (1): 153–182.

Daston, Lorraine, and Katharine Park. 1998. *Wonders and the Order of Nature*, 1150–1750. New York: Zone Books.

Droege, G., K. Barker, O. Seberg, J. Coddington, E. Benson, W. G. Berendsohn, B. Bunk, *et al*.2016. The Global Genome Biodiversity Network (GGBN) Data Standard Specification. Database 2016: baw125. https://doi.org/10.1093/database/baw125.

Dudley, Joseph P., and and Michael H.Woodford. 2002. Bioweapons, Biodiversity, and Ecocide: Potential Effects of Biological Weapons on Biological Diversity Bioweapon Disease Outbreaks Could Cause the Extinction of Endangered Wildlife Species, the Erosion of Genetic Diversity in Domesticated Plants and Animals, the Destruction of Traditional Human Livelihoods, and the Extirpation of Indigenous Cultures. *BioScience* 52 (7): 583–592.

Durkheim, Emile, and Marcel Mauss. 1963. *Primitive Classification*. Translated by R. Needham. Chicago: University of Chicago Press.

Ellen, R. 1993. *The Cultural Relations of Classification: An Analysis of Nuaulu Animal Categories from Central Seram*. Cambridge: Cambridge University Press.

Findlen, Paula. 1994. *Possessing Nature: Museums, Collecting, and Scientific Culture in Early Modern Italy*. Berkeley: University of California Press.

Findlen, Paula. 2002. Inventing Nature: Commerce, Art, and Science in the Early Modern Cabinet of Curiosities. In *Merchants & Marvels: Commerce, Science, and Art in Early Modern Europe*. New York: Routledge.

Foucault, Michel. 1966. *The Order of Things: An Archaeology of the Human Sciences*. New York: Vintage.

Genome 10K Project. 2016. Genome 10K Project. 2016. https://genome10k.soe.ucsc.edu.

Gosden, Chris, and Yvonne Marshall. 1999. The Cultural Biography of Objects. *World Archaeology* 31 (2): 169–178.

Greenblatt, Stephen. 1992. *Marvelous Possessions: The Wonder of the New World*. Oxford: Clarendon Press.

Haraway, Donna. 1997. *Modest_Witness@Second_Millennium.FemaleMan Meets⊗OncoMouse: Feminism and Technoscience*. New York: Routledge.

Hayden, Cori. 2003. From Market to Market: Bioprospecting's Idioms of Inclusion. *American Ethnologist* 30 (3): 359–371.

Henare, Amiria, Martin Holbraad, and Sari Wastell. 2007. *Thinking through Things: Theorising Artefacts Ethnographically*. London: Routledge.

Impey, Oliver, and Arthur MacGregor, eds. 1985. *The Origins of Museums: The Cabinet of Curiosities in Sixteenth-and Seventeenth-Century Europe*. Oxford: Clarendon Press.

Ingold, Tim. 2010. Bringing Things to Life: Creative Entanglements in a World of Materials. *World* 44 (July): 1–25.

Kay, L. E. 2000. *Who Wrote the Book of Life? A History of the Genetic Code*. Stanford: Stanford University Press.

Keller, Evelyn Fox. 2009. *The Century of the Gene*. Cambridge: Harvard University Press.

Kowal, Emma, Joanna Radin, and Jenny Reardon. 2013. Indigenous Body Parts, Mutating Temporalities, and the Half-Lives of Postcolonial Technoscience. *Social Studies of Science* 43 (4): 465–483.

Lampland, Martha, and Susan Leigh Star. 2009. *Standards and Their Stories: How Quantifying, Classifying, and Formalizing Practices Shape Everyday Life*. Ithaca: Cornell University Press.

Leonelli, Sabina. 2013. Why the Current Insistence on Open Access to Scientific Data? Big Data, Knowledge Production, and the Political Economy of Contemporary Biology. *Bulletin of Science, Technology & Society* 33 (1–2):6–11.

Leonelli, Sabina. 2014. What Difference Does Quantity Make? On the Epistemology of Big Data in Biology. *Big Data & Society* 1 (1).

Levi-Strauss, Claude. 1966. *The Savage Mind*. Chicago: University of Chicago Press.

Lowe, Celia. 2006. *Wild Profusion: Biodiversity Conservation in an Indonesian Archipelago*. Princeton: Princeton University Press.

Moutu, Andrew. 2006. Collection as a Way of Being. In *Thinking through Things: Theorising Artefacts Ethnographically*, edited by Amiria Henare, Martin Holbraad, and Sari Wastell. London: Routledge.

Mulcahy, Daniel G., Kenneth S. Macdonald, Seán G. Brady, Christopher Meyer, Katharine B. Barker, and Jonathan Coddington. 2016. Greater than X Kb: A Quantitative Assessment of Preservation Conditions on Genomic DNA Quality, and a Proposed Standard for Genome-Quality DNA. *PeerJ* 4 (October): e2528. https://doi.org/10.7717/peerj.2528.

Needham, R. 1979. *Symbolic Classification*. Santa Monica, CA: Goodyear.

Olmi, Giuseppe, Laura Laurencich-Minelli, Antonio Aimi, Vincenzo de Michele, Alessandro Morandotti, Elisabeth Scheicher, Rudolf Distelberger, Eliska Fucikova, Oleg Neverov, and Hans Christoph Ackermann. 2001. *The Origins of Museums: The Cabinet of Curiosities in Sixteenth-and Seventeenth-Century Europe*. London: House of Stratus.

Page, Lawrence M., Bruce J. MacFadden, Jose A. Fortes, Pamela S. Soltis, and Greg Riccardi. 2015. Digitization of Biodiversity Collections Reveals Biggest Data on Biodiversity. *BioScience* 65 (9): 841–842.

Parry, Bronwyn. 2004. *Trading the Genome: Investigating the Commodification of Bio-Information*. New York: Columbia University Press.

Poinar, Hendrik N., Carsten Schwarz, Ji Qi, Beth Shapiro, Ross D. E. MacPhee, Bernard Buigues, Alexei Tikhonov, *et al*.2006. Metagenomics to Paleogenomics: Large-Scale Sequencing of Mammoth DNA. *Science* 311 (5759): 392–394.

Pomian, Krystof. 1990. *Collectors and Curiosities: Paris and Venice, 1500–1800*. Cambridge: Polity Press.

Radin, Joanna. 2012. *Life on Ice: Frozen Blood and Biological Variation in a Genomic Age, 1950–2010*. Ph.D. Dissertation. Philadelphia: University of Pennsylvania.

Revive and Restore. 2013. Revive and Restore. 2013. http://longnow.org/revive.

Ridley, Matt. 2000. *Genome: The Autobiography of a Species in 23 Chapters*. New York: HarperCollins.

Rose, Nikolas. 2009. *The Politics of Life Itself: Biomedicine, Power, and Subjectivity in the Twenty-First Century*. Princeton and Oxford: Princeton University Press.

Shapiro, Beth. 2015. *How to Clone a Mammoth: The Science of de-Extinction*. Princeton: Princeton University Press.

Strasser, Bruno. 2012. Data-Driven Sciences: From Wonder Cabinets to Electronic Databases. *Studies in History and Philosophy of Science Part C: Studies in History and Philosophy of Biological and Biomedical Sciences* 43 (1): 85–87.

STRI. 2016. Smithsonian Tropical Research Institute. 2016. www.stri.si.edu.

Suarez, Andrew V., and Neil D. Tsutsui. 2004. The Value of Museum Collections for Research and Society. *BioScience* 54 (1): 66–68.

Sunder Rajan, Kaushik. 2006. *Biocapital: The Constitution of Postgenomic Life.* Durham, NC: Duke University Press.

Tsing, Anna. 2005. *Friction: An Ethnography of Global Connection.* Princeton: Princeton University Press.

Van Allen, Adrian. 2018. Pinning Beetles, Biobanking Futures: Practices of Archiving Life in a Time of Extinction. *New Genetics and Society* 37 (4): 387–410. https://doi.org/10.1080/14636778.2018.1546573.

Van Allen, Adrian. 2019. Folding Time: Practices of Preservation, Endangerment and Care in Making Bird Specimens. In *Deterritorializing the Future*, edited by Rodney Harrison and Colin Sterling. London: Open Humanities Press.

Zimov, Sergey A. 2005. Pleistocene Park: Return of the Mammoth's Ecosystem. *Science* 308 (5723): 796–798.

Zorach, Rebecca, Elizabeth Rodini, Sarah Cree, Alexandra M. Korey, Lia Markey, and Dawna Schuld. 2005. *Paper Museums: The Reproductive Print in Europe, 1500–1800.* Chicago: David and Alfred Smart Museum of Art, University of Chicago.

4 The kinship of bioinformation[1]

Relations in an evolving archive

Resto Cruz, Penny Tinkler and Laura Fenton

On 19 February 2020, the UK's Medical Research Council (MRC) Unit for Lifelong Health and Ageing announced on Twitter that they had been 'getting ready for the 74th Birthday mail out'.[2] The tweet includes a photo of staff members sitting inside a conference room in the Unit's new office in Central London. They are stuffing newsletters and birthday cards into window envelopes, piles of which are on the table. The intended recipients are the participants of the National Survey of Health and Development (NSHD), all born over the course of one week in March 1946. Since 1962, the Survey has been sending them birthday cards, the designs of which vary annually.[3] 'For many study members ours is the first card to drop through the letter box', the NSHD declares proudly on its website.[4] The Survey is the world's oldest continually running birth cohort study, and the first in the UK. To date, there have been 24 main waves of data generation.[5]

During its most recent sweep between 2014 and 2016, 2,214 men and women participated in the NSHD (Kuh et al. 2016). They initially received through the post a 32-page questionnaire. Composed of 70 main questions, topics covered ranged from family, home, and life since retirement, to health and hospital admissions, feelings and thoughts, lifestyle (including eating and sleeping habits), self-perception, and social life. A slightly smaller number of participants were subsequently visited at home by nurses who administered a hefty 55-page interview schedule. This time, questions focused on participants' medical conditions, everyday tasks and activities, health behaviour, reliance on personal help, and socioeconomic circumstances. Participants were also assessed: for their blood pressure, height, and weight, but also a range of physical and cognitive functions. In the middle of each visit, nurses gave participants a self-completion booklet on how they had been feeling the past four weeks. Where participants gave their consent, nurses drew blood samples. Finally, participants were asked if they would be willing to wear an activity monitor during the following week. The nurses left them with a further seven-page activity questionnaire.

In this chapter, we explore how birth cohort studies such as the NSHD register, and illuminate, the turn to bioinformation. With its emphasis on biological samples, measures, and processes, this turn appears to be radically different from postwar social science's concerns and methods (Savage 2010). Seen from the

DOI: 10.4324/9780367810030-4

perspective of the NSHD, however, the former builds on the latter, with relational work and infrastructure initiated earlier providing scaffolding for the contemporary generation of bioinformation.

We take as a point of departure Parry and Greenhough's description of bioinformational data as 'always relational', meaning that 'what data is depends on who uses it, how, and for what purpose' (2018: 7; emphasis removed). Their description resonates with wider discussions of how data is historically contingent – shaped, but not determined, by situated norms and expectations surrounding it (Bell 2015, Gitelman and Jackson 2013, Rosenberg 2018). Part of this relationality, too, are continuities and disjunctures between present and earlier forms of data (Lemov 2017). We see this relationality in birth cohort studies and epidemiology. Such studies enrol and generate heterogeneous (and in some cases, unrecognised) data; the form, creation, use, meaning, and implications of their data are shaped by the relations within which they are embedded (Bauer 2008, 2013, Kalender and Holmberg 2019).

Our account enriches this relational understanding of bioinformation by honing in on the long-term significance of kinship for how bioinformational data is generated and analysed, and how long-running cohort studies are maintained. Kinship, we suggest, is consequential for bioinformation, shaping both member participation and scientific analyses. While dominant understandings of kinship play a significant role in both, kinship (like data) is irreducible to such ideals. We build on work that foreground kinship's generativity and contemporary significance for biomedicine and science, while being mindful of its inherent complexity and difficulties (Carsten 2019a, 2019b, Das 2020, Opitz, Bergwik and Van Tiggelen 2016). Archives, such as that of the NSHD, illuminate the open-endedness and situatedness of scientific practices and data, and register the vitality and vicissitudes of kinship (Daston 2017, Garcia 2020).

We draw from an interdisciplinary project on girlhood in postwar Britain and its implications for later life.[6] As part of this project, we examined the NSHD and the English Longitudinal Study of Ageing as sources for understanding youth transitions and the life course for the generation of British women born between 1939 and 1952 (Tinkler, Cruz and Fenton 2021, Xue et al. 2020). We focus on our archival work with the NSHD, which entailed examining the accumulated records of 30 participants. We also draw from interviews with 30 other NSHD participants conducted in 2010 as part of a sub-study under the Healthy Ageing Across the Lifecourse (HALCyon) research programme (Elliott et al. 2011). We likewise engage publications produced by, and on, the NSHD. We bring to bear an ethnographic sensibility on these materials, paying attention to their dominant logics and the multiple voices and gaps they contain.

A birth cohort over time

The NSHD began as an instance of 'big social science': large-scale researches in the twentieth century that sought to document human life, at times aspiring for totality (Lemov 2017). Fuelled by concerns over declining fertility and maternal and infant health and mortality, it started in 1946 as a one-off survey of 13,687 babies born in Great Britain one week in March of that year (Royal College of Obstetricians and Gynaecologists and the Population Investigation Committee 1948). Realising the unique opportunity afforded for exploring postwar social changes – especially education reforms and the establishment of the National Health Service in 1948 – the Survey was reworked and extended. The NSHD was seen as a representative study of Great Britain, and it would play a pivotal part in policy debates surrounding schooling and health (Ramsden 2014). Participants were conceived as individuals who develop through time, and who belong to, and are shaped by, social groups (e.g. 'manual working' or 'middle' class). Understanding them was seen as a means of illuminating broader social groups and dynamics, including inequalities and the general state of Britain (Savage 2010: Chapter 8).

The initial sample was whittled down in 1948 to 5,362 singleton births within marriage; apart from losses due to attrition, these have remained the core of the NSHD. Survey members' mothers were the main participants in its early years – they answered postal questionnaires and were visited by municipal health workers. At age 13, members themselves started to answer questionnaires, and three years later, they became the study's main participants. Likewise, members' teachers, school nurses and doctors, and youth employment officers were asked to fill in questionnaires. Social science issues jostled with medical interests: childhood health, personality and mental health and wellbeing, education, environmental pollution, and entry into the labour force, among others. Questionnaires were wide ranging, included many open-ended questions, and 'rambled' (Pearson 2016: 111).

One reason for the NSHD's broad approach during this time is the ad-hoc nature of its funding, which required it to meet the priorities of different state agencies and charities (Ramsden 2014). In 1962, however, the MRC started to provide core funding, eventually transforming the Survey to a leading epidemiological study. The MRC established at the LSE the Unit for Research on the Environmental Background of Mental and Physical Illness, through which the NSHD continued to be run under the leadership of its founding director, James Douglas. Health became an important focus during this time, but Douglas himself continued to publish on education, occupation, and social mobility. To the extent that he analysed health outcomes, he emphasised environmental explanations, earning him some criticism from the MRC in 1968 for excluding genetics (Ramsden 2014: 140, note 64).

Following Douglas's retirement in 1979, the NSHD was downsized to an 'external scientific team' at the University of Bristol. Funding was uncertain, but Douglas's successor, Michael Wadsworth, convinced the MRC to give the Survey five years to prove its continued worth. In the meantime, Wadsworth led the collection of further data, including the functional tests carried out by nurses in 1982 mentioned above. The Bristol team was influenced by emerging areas of interest in epidemiology, including risk factors associated with cardiovascular diseases and the long-term effects of health-related behaviour. They also took advantage of the availability of new devices (e.g. random zero sphygmomanometer) and research nurses (Wadsworth 2014). Utilising their new data, NSHD researchers published a paper showing a relationship between low birthweight and high blood pressure in adulthood (Wadsworth et al. 1985). This finding anticipated what would become the 'Barker hypothesis', which looks at the foetal and developmental origins of adult diseases, and helped shift epidemiological focus away from adult lifestyles (Barker 1990, Kuh and Davey Smith 1993). Indeed, they beat Barker in publishing similar data, an achievement that features in Survey scientists' accounts of their history.[7]

The MRC hence decided to continue to fund the NSHD, which in 1987 moved to UCL where it remains to date. The late 1980s and early 1990s was a watershed period for the NSHD and UK birth cohort studies more generally, marking the 'end of a period of relative famine' (Kuh 2016: 1070). During this time, the Survey focused on the early origins of adult health and diseases (Wadsworth 1991). It has subsequently been at the forefront of life course epidemiology, which examines the later life health consequences of conditions and events throughout a person's – and their parents' – life (Kuh and Ben-Shlomo 1997). Amidst wider concerns regarding an ageing population, the Survey (along with two other smaller longitudinal studies) was embedded in 2008 in a new Unit, this time with a focus on Lifelong Health in Ageing.[8] It currently investigates biological ageing, including its relationship to mental and social wellbeing (Kuh 2016).

The data that the NSHD collects and the methods it employs have changed. Self-answered postal questionnaires remain important, but these are now populated mostly with close-ended questions: yes/no items, scales, and sections that require specific quantities. Since 1982, members have been visited by nurses who conduct tests, including those related to breathing, blood pressure, and physical function. Biological samples, including blood, saliva, urine, and cheek swabs, have been collected since 1999. Some have been used to extract DNA and biomarkers, with the rest stored for future analyses. In 2006, members visited research clinics for the first time, during which they underwent intensive functional tests and structural assessments of key body organs (Kuh et al. 2011). As mentioned, participants in the 2014–16 wave were invited to wear activity monitors. Since then, slightly over 500 participants have taken part in a substudy on dementia. These members visited clinical research facilities, where they underwent neuroimaging, cognitive testing, cardiovascular examination, and the collection of further biological samples (Lane et al. 2017, Mason et al. 2020). The Survey's plans include the use of wearable devices that track participants' physical activities, additional clinical visits, and continued linkage with health records.[9]

Hence, bioinformation – information generated from biological samples and which pertain to biological mechanisms and processes (Parry and Greenhough 2018) – have become fundamental to the NSHD. As in epidemiology and biomedicine more generally, this development entails 'big data': voluminous and complex datasets that require high computing power (Hoeyer et al. 2019). Crucial too is the rise of postgenomics in the life sciences, which departs from seeing the gene as stable (i.e. not modified by social and environmental factors) and determining of human biology and health, emphasising instead 'complexity, indeterminacy, and gene-environment interactions' (Stevens and Richardson 2015: 3). Here, the person is reframed from a unitary entity shaped by their social location into one that is made and remade in manifold relations that entail connections and disconnections, aggregation and disaggregation (Bauer 2013).

Indeed, the Survey (and the Unit) has positioned itself at 'the forefront of the big data revolution'. It seeks to harness innovations in biotechnology, imaging, and data science to shed light on the physiological, cardiovascular, metabolic, cognitive, and emotional aspects of ageing. Part of its ambition is to account for risk factors across the life course, thereby identifying those who are most at risk, and potentially, ways of intervening to improve health outcomes. By integrating novel bioinformational data and analytics with social and demographic data, the NSHD aims to illuminate the mechanisms through which social factors become inscribed in bodies, but also how those with similar experiences in life may have varying health outcomes.

Embracing big data and postgenomics comes with challenges, however. The Survey's historic significance, the finite nature of biological and genetic materials, and the need to pool resources to achieve analytical power, have meant increased demand for data access. Indeed, the Survey is increasingly part of cross-cohort studies and international consortia. There is pressure, too, from funders who are keen to maximise the high cost of maintaining birth cohort studies, especially with the integration of novel techniques (e.g. Medical Research Council 2014). Ownership and security are issues here, and the NSHD has introduced measures that govern data sharing.[10] These issues are pertinent to research collaborations in big data biology more generally (Pinel 2020), but heightened by the Survey's history of its own researchers exerting control over the study's data.[11] Meanwhile, the collection of biological samples and the use of clinical research facilities have led to apprehensions over members' continued participation, an issue revisited below. The planned incorporation of wearable devices has also generated concerns regarding privacy, including that manufacturers should not have access to the resulting data.[12] These concerns echo broader controversies over wearable devices and tracking applications and their use in cohort studies and epidemiological and health research.

There also anxieties surrounding the political implications of this ongoing turn. As alluded to, the Survey helped shape health and education policies during its early years. In more recent decades, it informed policies, such as those addressing disadvantages in early life, obesity, and health inequalities in later life.[13] Yet, there are concerns that new technologies might distract epidemiologists from questions of inequality, which have become less fashionable in the discipline (Kuh 2016: 1075). This more cautious view dovetails with critical scholarship on how developments such as postgenomics and big data could generate restricted understanding of socioeconomic inequalities, elide questions of historical injustices, and individualise health outcomes (e.g. Lock 2015). Moreover, this outlook is accompanied by a sense that the Survey's policy influence is under strain. This sense stems in part from the availability of other birth cohort studies that also shed light on health and the life course.[14] Relevant too is how life course epidemiology's findings cn the long-term factors behind health inequalities in later life do not square with current political realities in Britain (Kuh 2016: 1076).

Mothers and birthday cards

Funding, technological innovations, and shifts in scientific interest do not tell the whole story of bioinformation in the NSHD. Important too are the relationships that have allowed the Survey to survive over seven decades. We draw inspiration from those who have analysed the relational underpinnings of biomedicine and adjacent sciences. For example Das's (2020: Chapter 6) discussion of psychiatry in Delhi. Drawing from Foucault, she conceives of the family as a 'switch point' that mediates between the state, disciplinary institutions, and forms of relatedness. Although families might subject members to institutional power, they can generate 'residues' that allow different modalities of life to be created and inhabited. Scientific and biomedical institutions and scientific practices may also be suffused with forms and idioms of relatedness, or depend on actors, dynamics, and resources that are usually associated with the 'domestic' (Carsten 2019a, Opitz, Bergwik and Van Tiggelen 2016).

Recall that mothers were the main participants of the Survey until their children turned 16. As described above, the survey that would eventually become the NSHD was prompted by concerns over fertility and maternal and child health. Here, the domestic and the political entwined: discourses and policies pertaining to postwar nation- and state-building in Britain centred on motherhood (Davis 2013, Lewis 1992). During the NSHD's early years, health visitors and school nurses were asked to evaluate the cleanliness of the child, the bodily care provided by mothers, and the state of repair of the home. In subsequent years, mothers would provide information about their child's schooling, health, habits, interests, and activities.

For some Survey members, their mother's initial involvement generated a certain givenness to their own participation in the NSHD. For some too, there is acknowledgement of a generational obedience to, and respect for, health workers, nurses, doctors, and teachers, which made participation almost impossible to refuse – for mothers and children alike. A female participant from Scotland, for example, remarked, 'I don't even remember how it was that we all came to be. Did our parents have to agree to this? Was that in the first instance, do you think?'[15] As a former director made plain, recruitment and consent in 1946 was much simpler compared to the present: 'if someone was willing to see you, that was consent' and 'the response rates were over 90 percent probably because people didn't think they could choose not to participate' (Kuh, quoted in Pearson 2011: 24). Sheila Glass,[16] a participant born in the Northeast echoed these thoughts: 'I think the population was more biddable in those days', says Sheila. 'I never really asked her [my mum] about the survey. It's just always been part of my life' (Harrop 2017: 20).

The materials and interviews that we have examined pertain to those who have continued to participate in the NSHD until the 2010s. However, rates of successful contact were lowest when participants were between 16 and 35 years of age. Apart from the disruptions and mobilities associated with this period of the life course (see below), low contact rates have been attributed to members' new ability to choose to participate or not (Wadsworth et al. 2003). Importantly, not all mothers (or parents more generally) were cooperative or consistently so. In some cases, questionnaires bore remarks like 'This mother did not co-operate in the 1952 Survey but agrees to do so now'.[17] Some parents may have initiated their child's eventual withdrawal, perhaps 'because of anxiety about their child's capabilities or health, or because of a fear "that home visits would show up dissension in the family"' (Wadsworth 1992: 214, quoting Douglas 1976: 9). In adulthood, similar relational dynamics shape members' engagement with the Survey. Temporary refusals, for example, are often due to 'disturbance caused by separations, death of a relative or friend, house moving, and illness' (Wadsworth et al. 2003: 2196).

Participants' relationship with the NSHD was not only routed through their mothers/parents, but was directly cultivated by the Survey team over the decades. When the NSHD started sending cards in 1962 (the same year that it started to be under the MRC), members had started leaving school, in some cases moving elsewhere for work or studies, and eventually marriage. The cards are meant to encourage contact and continued participation, as they include requests for 'notification of changes of name and/or address, and now includes a review of recent work and references to publications' (Wadsworth et al. 2003: 2195). To capture the interest of participants, cards feature designs that are deemed appropriate to their age and gender (especially in the early years, when cards for boys were designed differently from those for girls), or which highlight important Survey milestones.[18] These are signed by current NSHD team members.

Many Survey members look forward to their birthday cards each year. One participant described phoning the NSHD office to ask about one card, which, for some reason, had been delayed.[19] Another shared how she would sometimes show the birthday card and the enclosed newsletter to friends and colleagues out of her own interest in how the Survey uses the information that it has collected over the years.[20] A participant from Southeast England noted how she often joked that 'it doesn't matter, I can fall out with all my family and all my friends. I can be the most awful person and I'll still get a card from the National Survey'.[21] This participant's statement signals both how cards are associated with (and are constitutive of) kinship and other close personal ties (di Leonardo 1987), and how the NSHD has used this association to ensure its continuity. Speaking to Pearson (2011: 24), one Survey member said, 'Somehow, over the years I began to feel I knew the team members, although I had never met any'.

The birthday cards have become especially important as the NSHD has increasingly incorporated home visits by nurses, clinical visits, and the collection of biological samples and measures. On various occasions, the Survey team expressed concern over the potentially adverse effects on participation of the NSHD's increased biomedical focus (Kuh et al. 2016, Wadsworth et al. 2003). To counteract these effects, the Survey has actively relied on sending birthday cards, and other practices, ranging from speaking to participants over the phone, to the occasional, but highly visible, media engagements:

> by being ever more diligent in responding to individual queries and comments, ensuring the experience of taking part is as smooth as possible and providing ever more tailored information and feedback; and by creating an even greater sense of identity with the study by requesting study members' personal experiences of being in the study, providing opportunities to meet other participants and the team through events held to mark the 65th and 70th birthdays, and working with them to increase the study's public profile.
>
> (Kuh 2016: 1073)

The practices described by Kuh are instances of the 'courtesy work' that Kalender and Holmberg (2019) argue is essential in ensuring both the participation of cohort members over an extended period of time and the quality of data that is generated. Building on this insight, we suggest that the NSHD's bioinformational turn occurred not only because of funding, the availability of technologies, and shifts in scientific interests, but also because of the relationship that the Survey has developed with its participants over seven decades. Notably, later birth cohort and longitudinal studies (including those whose focus have been biomedical from the outset) have adopted the sending of birthday cards and related practices to help maintain participation (Andersen, Madsen and Lawlor 2009: 117–18).

For some Survey members, participation allowed them to develop their sense of self (Tinkler, Cruz and Fenton 2021). Several of those interviewed in 2010 recalled feeling special when, as schoolchildren, they had to answer cognitive tests and questionnaires and undergo interviews. 'We were put in this class by ourselves for two or three hours. We used to think it was fantastic. All the rest of the class getting taught things and we were trying to fill these sheets out', a male participant from Scotland recalled. His brother, in fact, would feel jealous – something that the participant would evoke in later life:

> I was actually talking to my brother this morning about [the Survey]. I said, 'I'll need to go and get my housework done before this woman comes'. 'Oh, you've got a woman coming to the house?' 'I was just winding you up, because I'm a member of that—'. Because he used to be jealous because I was going and he wasn't.[22]

Some participants also referred to being able to contribute to society (see below). In some cases, though, they also referred to the medical examinations and clinical assessments as providing assurance and additional care that would not be easily available to them otherwise. A few likened these examinations to the annual MOT test for vehicles.

NSHD scientists take pride over the Survey's high contact and participation rates, given the general decline in participation in cohort and epidemiological studies (Kuh 2016, Kuh et al. 2016). The high rates also go against how ageing tends to decrease participation (Chatfield, Brayne and Matthews 2005, cited in Kuh 2016). In anticipation of possible cognitive decline among participants, the NSHD team has devised an incapacity protocol that would allow it to continue collecting data. More than 90 percent of participants agreed to the protocol, which was implemented in a small number of cases in the most recent wave (Kuh et al. 2016).

Death has also decreased the pool of participants in ways consistent with the national population (Wadsworth et al. 2003). The NSHD team anticipates that only some 300 members will make it to their 100th birthday (Pearson 2016: 304). Yet, for some Survey members, participation need not end at the moment of death. 'You're very aware that your memory is going... But you also know that in the archive is a version of you', one participant told Pearson (2011: 24). In some cases, the promise of immortality takes the form of the body itself. Here, we return to Sheila Glass, who, in an article written by Harrop (2017: 19) is quoted as saying: 'I've said they can have my brain and my body when I die'. She makes clear the value that she attaches to participating: 'If you can do something that might add to the sum total of mankind's knowledge. Then why not?' Her relationship with her mother, as discussed above, is part of the story, and so is her own relationship with the NSHD. We glean this when she quotes her husband who, perhaps in jest, but also in allusion to the tie between the Survey and its participants, asked: 'Does that mean we don't have to pay for a funeral?'

Comparisons and transmissions

Relations have also received considerable analytical attention from Survey researchers. They figure as sites for investigating intergenerational comparisons and transmissions. Generations are seen as distinct kinship layers that succeed each other over time, and which are connected by unidirectional and asymmetric flows (i.e. from older to younger) (Ingold 2007). Reflecting cultures of relatedness in twentieth-century Britain (Firth, Hubert and Forge 1969, Strathern 2011), the parent-child tie is of utmost significance, and so are marriage and the heteronormative nuclear family. As noted, notions of motherhood shaped the initial focus and design of the NSHD. Recall too that the Survey eventually focused on singleton births within marriages. This decision is justified on pragmatic grounds (Wadsworth 1991), but it is one that squares with prevailing moral norms in the postwar decades.

One key area that the NSHD investigated initially is how parental background and family life shaped schooling outcomes (Douglas 1964, Douglas, Ross and Simpson 1968). Data on parents' educational background and their interest in their child's schooling were collected. and members' attendance and academic performance were assessed. Analyses revealed how children's educational attainment was heavily influenced by their parents' social class. Parents' concern for their child's schooling improved the life chances of those coming from disadvantaged backgrounds, as did support from teachers. Protracted disruptions in family life (e.g. a parent's long-term illness) constrained children's achievements. These findings led to the establishment of non-selective comprehensive schools, and challenged prevailing views that explained educational achievement as the result of innate ability and heredity (Ramsden 2014).

In contrast with this emphasis on parents as sources of advantages and disadvantages, the implications of Survey members' lives for that of their parents have received scant attention. In 1989, at the age of 43, members were asked to complete a series of scale questions assessing the quality of their relationship with their parents during their childhood.[23] Were their parents (or parental figures) affectionate to them, or overprotective, for example? Members were also asked by nurses about the quality of their current relationship with their parents.[24] Papers written on the basis of these questions take members' relationship with their parents during their childhood as a factor that shapes subsequent outcomes, like mental wellbeing (e.g. Stafford et al. 2015). Less understood is how members' relationship to their parents in adulthood (e.g. caregiving) is shaped by earlier circumstances. To the extent that data on caregiving has been used, this has been in the context of examining factors that shape Survey members' own expectations for care provision in the future (Stafford and Kuh 2018).

To some extent, the NSHD has examined cohort members' relationship with their own children. From 1969 to 1975, the NSHD ran a 'second generation' study to investigate cohort members' parenting practices. As with the main study, mothers were key: female participants and the wives of male participants were interviewed when their firstborn child was four years old, and again when their child turned eight. Data from this study have been used to examine intergenerational continuities and discontinuities in cognitive ability and parenting practices (Byford, Kuh and Richards 2012, Wadsworth 1991). More recently, epidemiologists have drawn from the NSHD main study to assess how the timing of cohort members' marriage and parenthood, and their combination with work, have implications for members' health and wellbeing in later life (e.g. Lacey et al. 2015). Parent-child ties are not at the centre of analysis here, but we see a reversal of the usual direction of the relationship between parents and children – an exception that proves the rule.

Due to the emphasis on vertical kinship ties, lateral ties, such as siblingship, are relegated to the background. In the early years, siblings figured in relation to questions regarding household composition and domestic arrangements.[25] In 1961, when mothers were interviewed for the final time, they were asked to identify members' siblings' current occupation.[26] Although recorded by school nurses, this information was not coded and is only accessible through an examination of the questionnaires themselves. The only other time when questions specifically about siblings were asked was in 1972.[27] The NSHD is not unique in this case. Sociologists and historians working on the UK have only begun to examine sibling ties, usually in the context of childhood (Davidoff 2012, Davies 2014). Epidemiologists too have started to analyse siblingship and its capacity to illuminate causality across the life course (Strully and Mishra 2009).

In focusing on health and developing a life course approach, as well as incorporating clinical and biological measures, the NSHD has also investigated Survey members' biological inheritances from parents. While the NSHD did not collect biological materials from members' parents, it has data on their background, illnesses, and, as noted, their relationship with members. These data have been combined with recent biological measures to infer intergenerational aspects of health and wellbeing. Wulaningsih and colleagues (2018), for example, assess how father's and mother's age at birth of Survey members affect members' leukocyte telomere length (LTL) in later life. Considered a marker of biological ageing, the shortening of LTL is correlated with increased mortality risk. Using biological samples collected when members were age 53 (and for some, also at age 60 to 64), and accounting for parents' sociodemographic characteristics, the researchers confirmed a previously observed association between father's age of birth and child's LTL, but also identified a joint effect of parental ages.

Other studies track the biological implications of members' relationship with their parents (as well as parents' relationship with one another). Data on these relationships are assessed under the rubric of childhood or early life adversity, which includes parental separation, death, neglect, and abuse. Reflecting gendered assumptions surrounding childhood and parenthood, lack of or insufficient maternal bonding is likewise included here. Often, these relational difficulties are examined in relation to socioeconomic conditions during members' childhood. In one study, the implications of childhood adversities on DNA methylation are examined (Houtepen et al. 2018). DNA methylation (the addition of methyl groups to the DNA molecule) is an epigenetic process that marks ageing at the molecular level and is linked to age-related diseases, among others. Biological samples collected through cheek swabs from NSHD participants at age 53 were analysed along with those from a later birth cohort study. The authors found that certain methylation patterns across the two cohorts are associated with particular kinds of childhood adversity, suggesting that these experiences' effects persist until later in life. Studies such as this strive to trace how psychosocial and socioeconomic circumstances become translated to (or inscribed in) the biological – with epigenetic markers, for example, constituting a 'biosocial archive' (Relton, Hartwig and Davey Smith 2015).

The pictures of inheritance, gender, and kinship implied by postgenomics, big data biology, and epidemiology have received sustained attention from medical anthropologists, science studies scholars, and others (e.g. Lock 2015). Life course epidemiologists themselves have raised criticism. Some have accentuated the need to be circumspect when it comes to claims regarding postgenomic analyses' potential (e.g. Relton, Hartwig and Davey Smith 2015). Others have problematised the gendered assumptions that underlie particular kinds of epidemiological analysis of maternal influence on adult health and illness (Sharp, Lawlor and Richardson 2018). Still others have alluded to the limited definition of the 'family' that contemporary cohort studies have adopted – and intriguingly, to how future epidemiologists might find this surprising or unbelievable (Andersen, Madsen and Lawlor 2009: 102). Finally, some have cautioned against the predictive power of birth cohort and epidemiological studies in the face of randomness (Davey Smith 2011). To what extent these disputations will transform the NSHD and other similar birth cohort studies remains to be seen.

An archive of relations

In 2011, *Nature* published a feature article on the NSHD. It includes a half-page photo of the Survey's then director standing in front of red and black box files. The caption reads: 'Diana Kuh leads the UK National Survey of Health and Development, which has compiled thick files on more than 5,000 people since their birth in 1946'. Indeed, since its inception, the Survey has generated its own archive.

Similar to other scientific archives (Daston 2017: 5), the NSHD's is 'open-ended' and 'opportunistic', shapeshifting in relation to new theories, lines of inquiry, and technologies, as well as necessarily selective. It contains heterogeneous data that are subsequently assembled, including in unanticipated ways, to generate models and accounts of association and causation (Bauer 2008; 2013). Yet, archives can be used and read in different and unintended ways; they neither simply enact logics of domination nor render marginal subjects totally invisible (Chaudhuri, Katz and Perry 2010, Zeitlyn 2012). We thus turn to how the NSHD archive contains pictures of kinship that exceed how Survey scientists themselves have analysed relations. Using Das's (2020) description of the family as a 'switch point' discussed above, we attend to the relational 'residues' that have accrued over time.

We describe our approach in detail elsewhere (Tinkler, Cruz and Fenton 2021), including our openness to what constitutes data. Going beyond the qualitative analysis of coded data and select open-ended questions, we 'scavenge' questionnaires and include handwritten notes and marks on questionnaires, as well as letters and photos sent by participants. These elements have not been coded due to technological and resource constraints, but also because they have not been seen as sources of data. For example, postal questionnaires often have space for 'any other comments' to boost the impression that the Survey team are interested in members' views, a standard retention technique.[28] Even in its earlier years, the NSHD was not interested in the perspective and experiences of individual Survey members per se, but in the broader patterns that could be discerned (Ramsden 2014: 137). We use these scavenged materials to 'recompose' the biographies of select participants, paying attention to traces of participants' subjectivity, the relationships within which they are embedded, and the complex temporalities in their lives. Although we are interested in how their lives unfold over time, we also dwell on particular moments, and the futures and layers of history that they contain. We embrace the archive's gaps and interpretive possibilities.

Parents (especially mothers), as noted, are particularly prominent in the Survey, especially during its first 16 years. In some cases, this prominence translates to an overpowering presence in the lives of members. Here, we turn to one participant, who we have named Therese. One of several siblings, Therese was born in a North England town. Her parents were married not long prior to her birth. Lack of housing meant Therese had to live with relatives during her early childhood. During this time, her mother began to work intermittently as a typist, while her father found a job repairing radio and television sets, later setting up his own shop.

When Therese was a schoolgirl, her teachers considered her an 'average' student, noting at various points how she 'does not always work to full capacity', and how 'her attainment are [sic] mediocre'.[29] Reflecting the expansion of opportunities for girls in the postwar period, Therese entertained the possibility of a career in education or the creative arts. However, subsequent sweeps of the NSHD indicate that there was trouble at home. 'Her home background is unstable and difficult', her headmistress wrote economically at one point.[30] The following year, 1964 (Therese was now 18), the questionnaire sent to Therese's school is marked 'untraced'. It appears that she had left school in January of that year. She was deemed 'a most unsatisfactory student in every way'.[31]

Reading through Therese's records, feelings of puzzlement but also curiosity came to one of us (Resto): What could have compelled Therese to abandon her schooling despite having expressed aspirations earlier? What did her headmistress mean when she referred to Therese's home background? These questions obtained their answers once Resto scrolled down to the 1968 postal questionnaire. At this point, he encountered 22-year-old Therese residing some 40 miles away from her parents. Having married a few years earlier, she is now the mother of two, and is on the cusp of purchasing a house. Neither a teacher nor an artist, she is a fulltime housewife – 'precisely that', she scribbled below the answer that she had encircled. When asked if, given another chance, she would leave school at 17, she answered that she would choose to go to teacher training college after leaving school. Her explanation exceeded the allotted 1/3 of the page, requiring her to attach a slip of paper using Sellotape.

> I was forced to leave school for family reasons that seemed unavoidable at the time but in retrospect appears to be mainly selfishness on the part of my parents who had been leaning on me very heavily since the time of the collapse of my father's business shortly before my 'O' levels. It was impossible for [me] to study properly as I was having to take over the role of my mother because of a complicated pregnancy. Rather than reveal any

of this I allowed my behaviour at school to be explained as idles & it was
easy for my father to demand I leave schoo. when my mother was finally
taken to hospital about 6 weeks before my 'A' levels. At the same time I
was under emotional stress due to pressures exerted upon my fiancé by his
excessively overpossessive mother. These I solved by becoming pregnant
entirely by design. Machiavellian perhaps, but effective!! I then married
later in the year when my mother was finally able to cope. I ought to point
out that my parents' selfishness stems from the fact that I have an older
sister who had finished school and was wandering from job to job & at no
time was she asked to assist in any way.[32]

We see here how parental bonds could be suffocating and therefore propel
particular movements in Survey members' life course. Ties of relatedness could
be suffused with ambivalence, generative of hierarchies and inequalities, and
laced with the potential to fail (Carsten 2019b, Das 2020). Breaks in the fabric
of kinship could likewise be generative of possibilities (Garcia 2020). The con-
junction between Therese's father's financial troubles and her mother's pregnancy generated difficult demands, complicated further by her fiancé's own
troubled relationship with his mother. It appears that Therese neither actively
confronted her parents, nor did she choose to 'reveal' her 'impossible' situation
at the time. As she indicated, she obeyed her father. However, conforming to
one's elders may mask subtle and complex assertions of freedom and agency (Das
2018), as her becoming pregnant 'entirely by design' suggests. Although she was
only 18 at that time (and therefore below the then age of majority), her pregnancy
most likely compelled her parents to agree to her early marriage. Therese's explanation itself illustrates children's ability to hone their own voice and subjectivity –
and thus to critique their elders – using their cultural inheritances (in this case,
including participation in the Survey).

Although Therese escaped her parents' demands, the consequences of her
relationship with them implicated her own children. In the 'other comments'
section of the 1968 questionnaire, Therese wrote that she had been unprepared
for motherhood. Her older child, in particular, 'is very possessive towards me
and has a tendency to outburst of nervous anger', which she attributed to her
'confusion at being faced with something I was completely unprepared for'. She
added: 'It seems a pity though that my [older child] has had to suffer from the
mistakes I made on [them]'. Therese described, too, her desire to realise her
thwarted ambition once her children are old enough. Subsequent questionnaires
indicate that her marriage ended in the late 1960s, leading to years of separation
from her children, with consequences for her emotional and mental wellbeing. It
was not until after she remarried years later that her children came to live with
her. In the 1980s, she went to university, hence fulfilling her earlier ambition.

It is unclear if Therese's relationship with her parents improved over time –
her parents make few appearances in subsequent questionnaires – although the
limited material hints that this did not happen. In 1989, Therese reported that
she was emotionally 'not very close' to her parents – and had no contact with

them.[33] When asked to describe her relationship with her parents in her first 16 years, the overall picture that Therese provided is of distance and lack of warmth, support, and understanding. She was asked, too, if she felt that she was mistreated by her parents in any way, with space for an open-ended account. Therese's response is coded in the database as 'neglect/abuse'. However, her actual answer is no longer accessible. The 1989 questionnaires were destroyed in a fire in July 2006 before they could be scanned into PDFs.[34] Like other open-ended questions from other years, answers to this particular query were not encoded.

Therese's case attests to the cultural weight given to ties of filiation in British kinship. It is also suggestive of how generations overlap with, and affect, one another, thus complicating linear and unidirectional depictions (Ingold 2007: 116–19). However, the account that she gave in 1963 at 17 also points to how her relationship with her parents is entangled with her siblings. She mentioned her older sister who was 'wandering from job to job' as a reason behind her parents' demands. Therese's sister is older by two years, while Therese is older than her younger siblings by at least a decade. Therese did not provide details regarding her relationship with her siblings, but similar to another case that we encountered (Tinkler, Cruz, and Fenton 2021), her 1963 account indicates how dynamics among siblings may complicate vertical transmissions of resources and privilege.

One implication of the history of the NSHD's focus over the decades is that we are able to scavenge more data pertaining to relations in the questionnaires prior to 1982. During the Survey's first three decades, questionnaires were sent to participants more often. Moreover, the focus on family life, education, and employment seemingly encouraged at least some members to write about their kinship ties. Yet, even as the NSHD's focus became more biomedical, some cohort members continued to include notes about their personal and family circumstances. In some of the other cases that we have scavenged, participants wrote how particular relational dynamics or events have consequences for their health and wellbeing in later life – caring for elderly parents, the death of a grandchild, the end of a decades-long marriage, or in one case, getting married late in life. The 2010 interviews likewise suggest the consequences for health and wellbeing of relationships (Carpentieri and Elliott 2013).

Therese has continued to engage with the Survey's (now decreased) open-ended questions; she has also continued to write comments, even where these are not asked. In 2011, after reporting a medical condition, she makes it clear that her condition developed after 'a very stressful time of both bereavement and house moving'.[35] Although it is not entirely clear who Therese was grieving for, her response to another item in the questionnaire indicated that her mother had died several months earlier.

Conclusion

Parry and Greenhough's (2018: 7) description of bioinformation as 'always relational' emphasises the use of data. Taking this as our starting point, we have examined in this chapter the significance of kinship for the generation and analysis of bioinformation. Tracking the UK's NSHD, we have shown how the world's longest-running birth cohort study was routed, and has been maintained, through relations. Maternal ties are important here, but over the years, cohort members have also developed their own relationship with the Survey. Kinship ties themselves have been framed and analysed by the NSHD over the years in ways consistent with prevailing cultures of relatedness in twentieth-century UK. We see this in how the Survey has focused on the nuclear family defined by marriage, and how it has emphasised vertical parent-child ties marked by linearity, asymmetry, and unidirectionality. Yet, examining the NSHD archive illuminates how kinship exceeds the picture of relations privileged by the Survey. Through our practice of scavenging and recomposition, we have traced the significance of lateral ties, and the complexity and difficulty inherent in intergenerational relationships and kinship more broadly, including how children may critique their parents.

In rendering bioinformation as we have done here, we draw attention to how bioinformation is irreducible to the state, technological developments, and shifts in scientific convention. Important too are the relations that enable its collection, inform its analysis, and equally important, elude its grasp. Kinship here is a generative force for scientific inquiry (Carsten 2019a, Opitz, Bergwik and Van Tiggelen 2016), both in the 'big social science' era of the NSHD, and its contemporary turn to bioinformation. Following Das (2020), it acts as a 'switch point' that generates 'residues', traces of which are present in the archive, even as their significance is only beginning to be understood. Our account, moreover, makes clear that the shape and texture of kinship cannot be taken for granted: which ties are consequential, their meanings, and their affective character are shaped by personal biographies, familial histories, and broader social transformations.

Finally, critics have accentuated the narrowing of milieu that has accompanied particular forms of bioinformation (Lock 2015). Foregrounding how much bioinformation owes (practically, analytically) to kinship, while recognising the embeddedness of kinship within wider histories as we have done here is one response. How this emphasis on the relational character of bioinformation might enrich or transform the practice of epidemiology and life course research remains to be seen, especially in view of ongoing disputations within the field. How anthropologists, sociologists, historians, and other scholars of bioinformation interested in critical collaboration with epidemiologists and birth cohort scientists might grasp this opportunity is an open-ended question that, like the questionnaires we have examined, could elicit unexpected answers.

Notes

1 Funding for the research on which this chapter is based was received from the Economic and Social Research Council (ES/P00122X/1; PI: Penny Tinkler). We are grateful to colleagues at the MRC LHA for their time and assistance, particularly Andrew Wong, Maria Popham, Jane Johnson, Adam Moore, and Philip Curran.

Michael Wadsworth and Deborah Kuh generously shared their thoughts with us, for which we are very thankful. Our colleagues Baowen Xue and Anne McMunn prepared the sampling frame we used in selecting our cases. An earlier version of this chapter was presented during the 2019 American Anthropological Association conference in Vancouver, at the panel 'Towards an Anthropology of Bioinformation' organised by EJ Gonzalez-Polledo and Silvia Posocco. Noah Tamarkin's perceptive comments helped us revise our presentation into the present chapter. Some materials discussed here appeared first in Tinkler, Cruz, and Fenton 2021.

2 https://twitter.com/MRCLHA/status/1230072811068194816, accessed 10 May 2020.
3 https://issuu.com/nshd/docs/70-website-card-all, accessed 10 May 2020.
4 https://www.nshd.mrc.ac.uk/archive/birthday-card-gallery, accessed 10 May 2020.
5 Excluding sub-studies involving specific groups of participants and the recent cross-cohort Covid-19 study.
6 Transitions and Mobilities: Girls Growing Up in Britain 1954–76 and the Implications for Later-Life Experience and Identity. See: https://sites.manchester.ac.uk/girlhood-and-later-life.
7 Michael Wadsworth, interview by Resto Cruz and Laura Fenton, 2 September 2019, Bristol, United Kingdom; Diana Kuh, interview by Laura Fenton and Resto Cruz, 2 September 2019, Bristol, United Kingdom. See also Kuh 2016: 1070.
8 The two other studies are the Southall and Brent Revisited Study (SABRE) and LINKAGE-Camden, both of which are focused on particular districts of London. See: www.ucl.ac.uk/cardiovascular/research/population-science-and-experimental-medicine/mrc-unit-lifelong-health-and-ageing-ucl/about.
9 www.nshd.mrc.ac.uk/75th-birthday/talks-series/nshd-findings and www.nshd.mrc.ac.uk/75th-birthday/talks-series/clinics, accessed 19 March 2021.
10 See www.nshd.mrc.ac.uk/data/data-sharing.
11 As opposed, for example, to the UK's subsequent national birth cohort studies. Funded by the Economic and Social Research Council, the data sets of these later cohorts are deposited in the UK Data Service portal (www.ukdataservice.ac.uk).
12 www.nshd.mrc.ac.uk/75th-birthday/talks-series/nshd-findings, accessed 19 March 2021.
13 www.nshd.mrc.ac.uk/about-us/impact-nshd, accessed 19 March 2021.
14 This is based on Michael Wadsworth's account during the NSHD's 75th Birthday Talks, 9 March 2021. www.nshd.mrc.ac.uk/75th-birthday/talks-series/history-nshd, accessed 19 March 2021.
15 HALCYon Interview D08, 2010.
16 Harrop does not appear to use a pseudonym. We use pseudonyms for all other Survey members referred to in this chapter.
17 A.1 September 1952-January 1953 School Absence and Holiday Sickness Record, January 1953.
18 www.nshd.mrc.ac.uk/archive/birthday-card-gallery.
19 HALCYon Interview D05, 2010.
20 HALCYon Interview D06, 2010.
21 HALCYon Interview D23, 2010.
22 HALCYon Interview D01, 2010.
23 Form B Self-Completion, 1989.
24 Form A Main Questionnaire, 1989.
25 For example, AN3 March 1952 School Nurse's Interview with Mother.
26 A7 January 1961 Final Interview with Mother.
27 1972 Main Questionnaire Section A.
28 Michael Wadsworth, interview by Resto Cruz and Laura Fenton, 2 September 2019, Bristol, United Kingdom. See also Rich, Chojenta, and Loxton (2013).
29 S4a Teachers Questionnaire, July 1956; S7a Teachers Questionnaire, May 1959.
30 S11a School and College Questionnaire, October 1963.
31 S12 School and College Questionnaire, November 1964.

32 H3 Postal Questionnaire, March 1968.
33 Form A Main Questionnaire, 1989.
34 HL Deb, 2 November 2006, c47W. See: https://publications.parliament.uk/pa/ld199697/ldhansrd/pdvn/lds06/text/61102w0004.htm, accessed 30 June 2020.
35 2008–10 Postal Questionnaire, March 2008.

References

Andersen, Anne-Marie Nybo, Mia Madsen, and Debbie A.Lawlor. 2009. Birth Cohorts: A Resource for Life Course Studies. In Debbie A. Lawlor and Gita D. Mishra, eds., *Family Matters: Designing, Analysing and Understanding Family-Based Studies in Life Course Epidemiology*. Oxford: Oxford University Press, 99–127.

Barker, David. 1990. The Fetal and Infant Origins of Adult Disease. *British Medical Journal* 301: 1111.

Bauer, Susan. 2008. Mining Data, Gathering Variables and Recombining Information: The Flexible Architecture of Epidemiological Studies. *Studies in History and Philosophy of Biological and Biomedical Sciences* 39, 4: 415–428.

Bauer, Susan. 2013. Modeling Population Health: Reflections on the Performativity of Epidemiological Techniques in the Age of Genomics. *Medical Anthropology Quarterly* 27, 4: 510–530.

Bell, Genevieve. 2015. The Secret Life of Big Data. In Tom Boellstorff and Bill Maurer, eds., *Data, Now Bigger and Better!*Chicago: Prickly Paradigm Press, 7–26.

Byford, Michelle, Diana Kuh, and Marcus Richards. 2012. Parenting Practices and Intergenerational Associations in Cognitive Ability. *International Journal of Epidemiology* 41, 1: 263–272.

Carpentieri, J.D. and Jane Elliott. 2013. Understanding Healthy Ageing Using a Qualitative Approach: The Value of Narratives and Individual Biographies. In Diana Kuh*et al.*, eds., *A Life Course Approach to Healthy Ageing*. Oxford: Oxford University Press, 118–130.

Carsten, Janet. 2019a. *Blood Work: Life and Laboratories in Penang*. Durham, NC: Duke University Press.

Carsten, Janet. 2019b. The Stuff of Kinship. In Sandra Bamford, ed., *The Cambridge Handbook of Kinship*. Cambridge: Cambridge University Press, 133–150.

Chaudhuri, Nupur, Sherry J.Katz, and Mary Elizabeth Perry. 2010. Introduction. In Nupur Chaudhuri, Sherry J. Katz, and Mary Elizabeth Perry, eds., *Contesting Archives: Finding Women in the Sources*. Champaign, IL: University of Illinois Press, xiii–xxiv.

Das, Veena. 2018. On Singularity and the Event: Further Reflections on the Ordinary. In James Laidlaw, Barbara Bodenhorn, and Martin Holbraad, eds., *Recovering the Human Subject: Freedom, Creativity and Decision*. Cambridge: Cambridge University Press, 53–73.

Das, Veena. 2020. *Textures of the Ordinary: Doing Anthropology After Wittgenstein*. New York: Fordham University Press.

Daston, Lorraine. 2017. Introduction: Third Nature. In Lorraine Daston, ed., *Science in the Archives: Pasts, Presents, Futures*. Chicago: University of Chicago Press, 1–14.

Davey Smith, George. 2011. Epidemiology, Epigenetics and the 'Gloomy Prospect': Embracing Randomness in Population Health Research and Practice. *International Journal of Epidemiology* 40, 3: 537–562.

Davidoff, Leonore. 2012. *Thicker Than Water: Siblings and their Relations 1780–1920*. Oxford: Oxford University Press.

Davies, Katherine. 2014. Siblings, Stories and the Self: The Sociological Significance of Young People's Sibling Relationships. *Sociology* 49, 4: 679–695.

Davis, Angela. 2013. *Modern Motherhood: Women and Family in England, 1945–2000.* Manchester: Manchester University Press.

Di Leonardo, Michaela. 1987. The Female World of Cards and Holidays: Women, Families, and the Work of Kinship. *Signs* 12, 3: 440–453.

Douglas, James W.B. 1964. *The Home and the School.* London: MacGibbon and Kee.

Douglas, James W.B. 1976. The Use and Abuse of National Cohorts. In Marten Shipman, ed., *The Organisation and Impact of Social Research*, 3–21. London: Routledge.

Douglas, James W.B., J.M. Ross, and H.R. Simpson. 1968. *All Our Future: A Longitudinal Study of Secondary Education.* London: Peter Davies.

Elliott, Jane, Catharine Gale, Diana Kuh, and Samantha Parsons. 2011. *The Design and Content of the HALCyon Qualitative Study: A Qualitative Sub-study of the National Survey of Health and Development and the Hertfordshire Cohort Study.* CLS Cohort Studies Working Paper 2011/5, Centre for Longitudinal Studies, Institute of Education, London.

Firth, Raymond, Jane Hubert, and Anthony Forge. 1969. *Families and their Relatives: Kinship in a Middle-Class Sector of London.* London: Routledge and Kegan Paul.

Garcia, Angela. 2020. Fragments of Relatedness: Writing, Archiving, and the Vicissitudes of Kinship. *Ethnos* 85, 4: 717–729.

Gitelman, Lisa and Virginia Jackson. 2013. Introduction. In Lisa Gitelman and Virginia Jackson, eds., *'Raw Data' Is an Oxymoron.* Cambridge, MA: MIT Press, 1–14.

Harrop, Sarah. 2017. 71 Years Under the Lens of Medical Research. *MRC Network,* September.

Houtepen, Lotte C., Rebecca Hardy, JaneMaddock, *et al.*2018. Childhood Adversity and DNA Methylation in Two Population-Based Cohorts. *Translational Psychiatry* 8: 266.

Hoeyer, Klaus, Susan Bauer, and Martyn Pickersgill. 2019. Datafication and Accountability in Public Health: Introduction to a Special Issue. *Social Studies of Science* 49, 4: 459–475.

Ingold, Tim. 2007. *Lines: A Brief History.* London: Routledge.

Kalender, Ute and Christine Holmberg. 2019. Courtesy Work: Care Practices for Quality Assurance in a Cohort Study. *Social Studies of Science* 49, 4: 583–604.

Kuh, Diana. 2016. From Paediatrics to Geriatrics: A Life Course Perspective on the MRC National Survey of Health and Development. *European Journal of Epidemiology* 31, 11: 1069–1079.

Kuh, Diana and Yoav Ben-Shlomo, eds. 1997. *A Life Course Approach to Chronic Disease Epidemiology.* Oxford: Oxford University Press.

Kuh, Diana and George Davey Smith. 1993. When is Mortality Risk Determined? Historical Insights into a Current Debate. *Social History of Medicine* 6, 1: 101–123.

Kuh, Diana, Mary Pierce, Judith Adams, *et al.*2011. Cohort Profile: Updating the Cohort Profile for the MRC National Survey of Health and Development: A New Clinic-Based Data Collection for Ageing Research. *International Journal of Epidemiology* 40, 1: e1–9.

Kuh, Diana, Andrew Wong, Imran Shah, *et al.*2016. The MRC National Survey of Health and Development Reaches Age 70: Maintaining Participation at Older Ages in a Birth Cohort Study. *European Journal of Epidemiology* 31, 11: 1135–1147.

Lacey, Rebecca, Mai Stafford, Amanda Sacker, and Anne McMunn. 2015. Work-Family Life Courses and Subjective Wellbeing in the MRC National Survey of Health and Development (the 1946 British Birth Cohort Study). *Population Ageing* 9, 1–2:69–89.

Lane, Christopher, Thomas D. Parker, Dave M. Cash, et al. 2017. Study Protocol: Insight 46 – A Neuroscience Sub-study of the MRC National Survey of Health and Development. *BMC Neurology* 17: 75. https://doi.org/10.1186/s12883-017-0846-x.

Lemov, Rebecca. 2017. Anthropology's Most Documented Man, Ca. 1947: A Prefiguration of Big Data from the Big Social Science Era. *Osiris* 32: 21–42.

Lewis, Jane E. 1992. *Women in Britain Since 1945: Women, Family, Work, and the State in the Post-War Years*. Oxford: Blackwell.

Lock, Margaret. 2015. Comprehending the Body in the Era of the Epigenome. *Current Anthropology* 56, 2: 151–177.

Mason, Sarah Ann, Lamia Al Saikhan, Siana Jones, *et al.*2020. Study Protocol – Insight 46 Cardiovascular: A Substudy of the MRC National Survey of Health and Development. *Artery Research* 26, 3: 170–179.

Medical Research Council. 2014. *Maximising the Value of UK Population Cohorts: MRC Strategic Review of the Largest UK Population Cohort Studies*. London: Medical Research Council.

Opitz, Donald L., Staffan Bergwik, and Brigitte Van Tiggelen. 2016. Introduction: Domesticity and the Historiography of Science. In Donald L. Opitz, Staffan Bergwik, and Brigitte Van Tiggelen, eds., *Domesticity in the Making of Modern Science*. Basingstoke: Palgrave Macmillan, 1–15.

Parry, Bronwyn and Beth Greenhough. 2018. *Bioinformation*. Cambridge: Polity.

Pearson, Helen. 2011. Study of a Lifetime. *Nature* 471: 20–24.

Pearson, Helen. 2016. *The Life Project*. Harmondsworth: Penguin.

Pinel, Clémence. 2020. When More Data Means Better Results: Abundance and Scarcity in Research Collaborations in Epigenetics. *Social Science Information* 59, 1: 35–58.

Ramsden, Edmund. 2014. Surveying the Meritocracy: The Problems of Intelligence and Mobility in the Studies of the Population Investigation Committee. *Studies in History and Philosophy of Biological and Biomedical Sciences* 47: 130–141.

Relton, Caroline L., Fernando Pires Hartwig, and George Davey Smith. 2015. From Stem Cells to the Law Courts: DNA Methylation, the Forensic Epigenome and the Possibility of a Biosocial Archive. *International Journal of Epidemiology* 44, 4: 1083–1093.

Rich, Jane Louise, Catherine Chojenta, and Deborah Loxton. 2013. Quality, Rigour and Usefulness of Free-Text Comments Collected by a Large Population Based Longitudinal Study – ALSWH. *PLOS One* 8, 7: e68832. https://doi.org/10.1371/journal.pone.0068832.

Rosenberg, Daniel. 2018. Data as Word. *Historical Studies in the Natural Sciences* 48, 5: 557–567.

Royal College of Obstetricians and Gynaecologists and the Population Investigation Committee. 1948. *Maternity in Great Britain*. Oxford: Oxford University Press.

Savage, Mike. 2010. *Identities and Social Change in Britain Since 1940: The Politics of Method*. Oxford: Oxford University Press.

Sharp, Gemma C., Deborah A.Lawlor, and Sarah S. Richardson. 2018. It's the Mother!: How Assumptions About the Causal Primacy of Maternal Effects Influence Research on the Developmental Origins of Health and Disease. *Social Science and Medicine* 213: 20–27.

Stafford, Mai, Catharine R.Gale, GitaMishra*et al.*2015. Childhood Environment and Mental Wellbeing at Age 60–64 Years: Prospective Evidence From the MRC National Survey of Health and Development. *PLoS One* 10, 6: e126683–126612.

Stafford, Mai, and Diana L. Kuh. 2018. Expectations for Future Care Provision in a Population-Based Cohort of Baby-Boomers. *Maturitas* 116: 116–122.

Stevens, Hallam and Sarah S. Richardson. 2015. Beyond the Genome. In Sarah S. Richardson and Hallam Stevens, eds., *Postgenomics: Perspectives on Biology After the Genome*. Durham, NC: Duke University Press, 1–8.

Strathern, Marilyn. 2011. What is a Parent? *HAU: Journal of Ethnographic Theory* 1, 1: 245–278.

Strully, Kate W. and Gita D. Mishra. 2009. Theoretical Underpinning For the Use of Sibling Studies in Life Course Epidemiology. In Debbie A. Lawlor and Gita D. Mishra, eds., *Family Matters: Designing, Analysing and Understanding Family-Based Studies in Life Course Epidemiology*. Oxford: Oxford University Press, 39–56.

Tinkler, Penny, Resto Cruz, and Laura Fenton. 2021. Recomposing Persons: Scavenging and Storytelling in a Birth Cohort Archive. *History of the Human Sciences*. Online first. https://doi.org/10.1177/0952695121995398.

Wadsworth, M. E. J. 1991. *The Imprint of Time: Childhood, History and Adult Life*. Oxford: Clarendon Press.

Wadsworth, M. E. J. 2014. Focussing and Funding a Birth Cohort Study Over 20 Years: The British 1946 National Birth Cohort Study From 16 to 36 Years. *Longitudinal and Life Course Studies* 5, 1: 79–92.

Wadsworth, M. E. J., H. A. Cripps, R. A. Midwinter, and J. R. T. Colley. 1985. Blood Pressure at Age 36 Years and Social and Familial Factors, Cigarette Smoking and Body Mass in a National Birth Cohort. *British Medical Journal* 291: 1534–1538.

Wadsworth, M. E. J., S. L. Butterworth, R. J. Hardy*et al.*2003. The Life Course Prospective Design: An Example of Benefits and Problems Associated with Study Longevity. *Social Science and Medicine* 57, 11: 2193–2205.

Wulaningsih, Wahyu, Rebecca Hardy, Andrew Wong, and Diana L. Kuh. 2018. Parental Age and Offspring Leukocyte Telomere Length and Attrition in Midlife: Evidence From the 1946 British Birth Cohort. *Experimental Gerontology* 112: 92–96.

Xue, Baowen, Penny Tinkler, PaolaZaninotto, and Anne McMunn. 2020. Girls' Transition to Adulthood and Their Later Life Socioeconomic Attainment: Findings From the English Longitudinal Study of Ageing. *Advances in Life Course Research*. Online first. https://doi.org/10.1016/j.alcr.2020.100352.

Zeitlyn, David. 2012. Anthropology in and of the Archives: Possible Futures and Contingent Pasts. Archives as Anthropological Surrogates. *Annual Review of Anthropology* 41: 461–480.

5 Bioinformation in formation

Inventing medical devices in contemporary India

Anisha Chadha

I am in an Indian Institute of Technology (IIT) innovation center, facing a scaffolding that spans the entire west wall of the "conceptualization room." Royal blue drawers are stacked in every groove of the scaffolding, less than an inch apart. I count from floor to ceiling, wall to wall, and multiply. Four hundred and two drawers, each with a small white label. Stepping closer to read the tags, I see that each drawer contains unique components. As this is an exceptionally extensive inventing space, all the usual materials are present: screws, rubber bands, tape, metal sheets, cables, plastic pipe, magnets, clay.[1] One of the inventors guides me from the raw materials room into an area devoted to creating virtual devices, then to a tiny room where they experiment with plastics and 3D print models. Beyond these alcoves, we climb half a flight of stairs to a makeshift loft where I meet the inventors who mock-up electronic devices. After brainstorming ideas in the conceptualization room, all the engineers are free to scatter across the innovation center, tinkering with raw components to assemble rough models.

Amidst the tens of multistory buildings in this sprawling IIT campus, the center feels like an odd cross between a hardware store, a machine shop, and a children's treehouse, a place for a small, select club to experiment, craft and play.[2] Across the entrepreneurial sector, South Asian inventors reason that surrounding themselves with as many raw materials as possible will result in the most creative designs.[3] And over my years of ethnographic research examining medical device production in India—across startups, tech incubators, and academic institutions like this IIT center—I have become used to the range of cramped, cluttered spaces in which inventors gather to brainstorm and build new biomedical technologies. In this chapter, I focus on such entrepreneurial teams in contemporary India to explore how bioinformation is gathered, produced, and animated via the creation of novel Indian medical devices, or "medtech."

Reflecting on ethnographic fieldwork in high-tech sites in India—primarily Bangalore, Delhi, Bombay and Chennai—I detail how medtech entrepreneurs generate a series of preliminary iterations, or prototypes, of each device. Prototyping is a key early stage in any novel technology's creation, one that requires the continuous production and feedback of human bioinformation. Ideally, each prototype will be a slightly improved version of the one prior, as

DOI: 10.4324/9780367810030-5

inventors progress toward their aim of eventually constructing something efficacious and marketable. Regardless of the clinical specialty for which it is being designed, nearly every medtech prototype depends on high proximity to human bodies from its earliest moments of materialization. Once engineers begin to interact with material objects, a range of functional, ergonomic, and accuracy issues that were previously invisible are made visible. For example, prosthetics are adjusted as patients demonstrate preferences for mobility, skin tone or bulkiness; diagnostic kits are altered until readings for blood sugar, pulse, temperature and vital signs can be trusted to match clinical standards; digital apps are tweaked until the back-end coding effectively corresponds to doctors' visual and tactile preferences for front-end screen options.

My insights into an anthropology of bioinformation stems from the close observations of such early moments of device materialization, engaging the technical, social and ethical dilemmas that present when the emergence of a new biomedical technology depends on human bio-feedback, before documented trials and widespread hospital uptake begin. Rather, the invention stages preceding the circulation of any device in progress beyond its small group of creators, as a finalized, identifiable object, are critical to its potential success. I argue that the necessary, continuous feedback of such data back into the prototyping process reveals bioinformation to be a key component on which the successful *creation* of future technologies also depends. Further, this attention to the continual feedback of data drawn from day-to-day human responses to medtech prototypes also reveals how bioinformation creates new human subjectivities in this South Asian context.

In what follows, I will first contextualize my research sites, asking what scholarly engagements with the prototype concept can offer an anthropology of bioinformation. Then, I will detail an ethnographic scene in order to engage theoretical, ethical, and methodological discussions about experimental test subjects. I will conclude with remarks about the ways bioinformation is co-producing a variety of new human subjectivities, notably test patients and citizen-entrepreneurs, alongside novel tech objects in the Indian medtech sector. Several emergent Indian medtech products certainly do elicit, record and sell biometric data from patient populations. Indeed, many devices procured by commercial manufacturers are invented for this exact purpose. Yet, my overall argument in this chapter is that tracing the processes and actors involved in very early-stage prototyping of these devices—before they are mass marketed—complicates the idea that bioinformation is simply data that technologies extract from bodies. Moving beyond this linear formula, the case of Indian medtech innovation demonstrates how variable inputs and outputs of bioinformation make possible the provisional relations between data and devices, and inventors and patients, required during early-stage technology production. In this context, bioinformation is more accurately perceived as a set of semiotic-material relationships (Haraway 1988), that yield not only finalized devices from prototypes, but also the production of new social identities of makers and users of biotech in contemporary India.

Sites of experimentation

Over the past decade, in globally connected high-tech sites, increasing numbers of doctors, engineers, and designers are redefining themselves as entrepreneurs and forming small teams to develop medical devices (Chaturvedi 2015, Ogrodnik 2012, Yock et al. 2015). In contrast to biotech—a term which traditionally encompasses genomics and pharmaceutical innovation—medtech refers to the creation of all other material technologies used in clinical settings. The demarcation between biotech and medtech can be reduced, respectively, to drug versus device creation. This distinction is routinely accepted across high-tech urban sites. India, a nation long recognized for engineering and technology expertise, has cultivated a robust medtech ecosystem since the late 2000s. This period has been marked by the return migration of many elite members of the diaspora (Bhatt 2018); the current right-wing BJP administration investing in citizen hackers and funding entrepreneurial training programs across top public universities and hospitals (Irani 2019); as well as the widespread trend of young STEM professionals leaving traditional career paths to work in startup companies (Goyal 2019).

Consequently, hundreds of doctors, engineers and designers currently form startups across Indian urban hubs to create novel devices intended to tackle a range of health issues. Their resulting technologies are being adapted to various clinical settings across the South Asian subcontinent and beyond. Indian medtech inventors are creating an assortment of material objects, from "low tech" stents, endoscopes, prosthetics and surgical equipment; to digital smartphone apps and wearables that monitor patients; to AI devices that track vital signs and neonatal development. Additionally, they are designing various testing kits, surgical equipment, and diagnostic tools. Seeking to invent across infrastructural and resource constraints—for public and private hospitals, and rural and community settings—entrepreneurs overwhelmingly cite a calling to innovate in India and for India as their driving motivation.

Tracing the *production* of medical devices also provides an understanding of the ways cultures of innovation embed existing power imbalances between entrepreneurs and their target patient populations and markets into new forms of South Asian biocapital (Sunder Rajan 2009). Following invention practices also exposes the range of human bodies, including the inventors' own, with which medtech entrepreneurs require their prototypes to engage in informal, undocumented relationships. Working as an ethnographer in their innovation spaces, I sometimes simultaneously offered my own body as a site of prototype experimentation, as well as a vehicle for laboratory and clinical participant observation. For example, over months of prototyping a diabetes screening kit with engineers and designers in Bangalore, team members often took turns performing a 12-hour fast in order to stop at local doctor's office to obtain a quick bloodwork panel in the mornings before work. We could then compare these figures to the output of the prototype being designed by pricking each other's fingers during the workday. These hasty, makeshift biometric checks afforded valuable data comparisons between monthly device in progress testing conducted with a women's group in a nearby *Marathi* urban slum community. Such shifting constellations of visceral engagements between actors involved in the prototyping process blurs boundaries between experimental subject, tech developer, and ethnographer. As debates about the benefits and dangers of tech industries' nonconsensual surveillance of consumers intensify, the physical closeness of hardware tinkering and healthcare delivery make the medtech entrepreneurial space an exceptionally stimulating niche of the global Silicon Valley landscape in which to investigate bioinformation from anthropological perspectives.

By examining the iterative nature of the invention process itself, investing my research efforts in examining the painstaking time of trial-and-error, before devices gain a solidified ontological presence and referential identity in the world, I used classic actor-network methods to follow prototypes as they are constructed from raw materials (Latour 1988). This approach allows me to materially trace the co-production of new biocapital with previously unrecognized experimental subjects, with emerging inventor subjectivities, and of course, with newly produced data. Healthcare prototypes need to be tested on human bodies to become clinically viable. The data that results from such experimentation is potentially indicative of patient wellness, but they also serve as "vital signs" in the invention process, where they indicate whether the technology-in-progress needs to be tweaked further or not. Medtech invention thus produces a cascade of bioinformation from which new Indian biocapital—in the form of technologies, entrepreneurs and patients—takes shape.

Proto-medtech actors

In Greek mythology, Proteus is a sea deity who possesses dual powers to foretell the future and mutate his shape. Stemming from this etymology, protean has come to mean that which is versatile, incomplete and transfigurable. Scientific nomenclature is rife with the proto- prefix. Protozoa, for example, is a taxonomic kingdom that encompasses the earliest multi-celled organisms. Paradoxically, however, proto- was only assigned as a bacteria phylum after these microorganisms were understood to be highly mutable life forms from which others evolved (Woese 1987). Across such disciplinary and popular understandings, the proto- qualifier now appears before a range of potential, yet unfinished, things. Proto-things are precursors to their future versions. They are recognizable, yet different, from their eventual, final forms.

Engineers and designers apply the prefix proto- to several dimensions of the physical world with which humans engage. The prototype concept is central to their theories and practices of building material models across a range of enterprises. Once initial prototypes are sketched and mocked up from available raw materials, they require an unpredictable number of tweaks to arrive at the last stage of the design process. Medtech inventors whose products have already entered the market created anywhere between four to two hundred prototypes en route to those completed devices. Medtech prototypes are often initially built by hand, using computer software or common shop machinery like lathe machines, 3D printers, and drill presses. In the previously mentioned IIT invention center, for instance, inventors routinely wrote code sequences to print surgical dental implants; experimented with metal bolts and sheets of plastic to create prosthetic legs; and pulled resistors and capacitors from the blue drawers to plug into electronic circuit boards. Once a prototype is mocked up, inventors need to be able to quickly verify that it is more accurate and effective than the previous version, altering experimental practices and device design in response to this bioinformation.

The most convenient data feedback inventors rely upon is to simply test the prototype against their own bodies, or those who happen to be proximal to their workspaces. This feedback loop of inductive reasoning contrasts historic—often quite mythic—notions of systematic, deductive scientific and engineering protocols. Yet the process of designing any new material technology requires a great number of iterative stages, as problems in functionality, production and marketability only reveal themselves in real-time through tests against human tech users. Prototypes thus have a perpetually unfinished material quality, and inventors do not invest effort and resources into the "finish" of a new machine or device until they deem to have arrived at the final version. Studies of hardware production and use have illustrated how the success of digital material technologies is measured by their ability to unambiguously depict and predict "immaterial behavior" (Kirschenbaum 2008). Anthropological attention to cultures of tech production, however, lends insight into which immaterial behaviors endow these practices with meaning that allow engineers and designers to construct marketable devices from raw hardware components in the first place. Ethnographically tracing the creation of biomedical technologies in particular yields great insight into these semiotic-material exchanges, as medtech prototyping inherently relies on the collection and analysis of bodily data.

George Marcus suggests that anthropology as a discipline should adopt engineers and designers' notion of prototyping, as it denotes an experimental mode analogous to the iterative nature of ethnographic fieldwork. The prototype, according to Marcus, is a "set of concepts in material form, far from advanced in development, but still open to revision, experiment, and some rethinking, based, in part, on engagement with 'others'" (2014). Like in medtech invention, the majority of the banal moments in the long phases of ethnographic research are later not deemed useful or imperative for the creation of the final product. Yet select bits of information from these visceral engagements still hold importance, as they could be amplified into larger structures and meanings in varying future times and places. The *initial processes*—of both ethnography and prototyping—are valuable for understanding how and why bioinformation that might later end up embedded in devices, archives, datasets and publications came to be collectively regarded as notable. Thus, biomedical technologies provide yet another example of the growing confluence of ethnographic methods in academic and industry professions, notably in design fields (see Murphy 2016). Continuing the analogy, I suggest that Indian medtech prototypes require unacknowledged, yet foundational, human expressions, adjustments and experiments in order to come into being.

Engaging with the iterative moments of embodied trial-and-error that prototyping requires, therefore, takes on a dual valence for ethnographers interested in scientific and engineering experimental practices. Translating my research objectives to my entrepreneur interlocutors, being co-present with them as we tested technologies on our and others' bodies, and systematically recording these moments as small bits of ethnographic data, allowed me to lay claim to what is effective, ethical and of interest to others. Further, reflexive auto-ethnography provides another instance of bioinformation being used to turn something proto- into a final product. As anthropologists studying bioinformation, how might we take into account that we are also taking data from bodies, or rather, co-producing data and subjects as a result of ethnographic relationships?

Kalindi Vora, in her studies of biopolitics and labor in contemporary India, expands the focus of Indian biocapital as more than "the economy of biological materials, production, and value" (2015: 4). Instead, she urges scholars to go beyond traditional ideas of materiality and take into account gendered and racialized histories that have led to the ways South Asians participate in global economic and cultural assemblages today. She writes that "the notion of life animating this understanding of biocapital references human biological and social existence as inseparable" (ibid). Vora's conceptualization of biocapital as animated by "vital energy"—defined as valorized investments of human energy in individual and social bodies—helps us understand how bioinformation is produced in the *relations between* prototypes, devices, and data, as well as between inventors, patients and anthropologists. Medtech devices can only output biometric data with human bodies as referents. And these devices are produced from a series of prototypes that are constructed within a field of semiotic and bio-material exchanges. Medtech producers in India view themselves as an interconnected entrepreneurial community. Longstanding personal relationships bridge divides within and between companies; "vital energy," between individual entrepreneurs and the community as a whole, inform how they produce and circulate novel biocapital.

As stated earlier, medical device innovation in India over the past decade increasingly occurs within small startup teams. Along with the co-presence required for health technologies to measure and track bodily outputs, building and testing medtech prototypes thus involves a great deal of physical and social intimacy. Startups generally consist of three to ten entrepreneurs, who are most often elite, high-caste, privileged men. Compared to technologies in other Indian startup niches—fintech, infotech, etc.—medtech devices take longer to invent. Prototyping and clinical trials require longitudinal human subjects data, and regulatory approvals must be procured before the device is eventually marketed. Medtech startup teams, therefore, often work closely for five to ten years on particular devices, developing very close friendships in the process. In fact, entrepreneurs who are able to eventually bring devices to market often attribute their success to the social dynamics of the team over their technical prowess.

Despite media portrayals of Silicon Valley technologists' life as full of "ah ha" moments and genius breakthroughs, anthropologists have demonstrated that high-tech culture is rife with mundane tasks and everyday relationships (English-Lueck 2002, Shankar 2008). In medtech, product designers are required to gauge how potential patients might react to micro elements of the device. Every time test subjects encounter the new version of the prototype, the entrepreneurs often make minute tweaks to colors, fonts and dimensions of the device, sometimes over tens or hundreds of repetitions. For the aforementioned diabetes screening kit, for instance, designers varied the spatial order of components of the kit countless times, meticulously documenting every slight light alteration in this "workflow"—the specific spatiotemporal sequence for which the device is designed to be effective. For example, objects placed millimeters apart on the testing table, specifications of blood drawn, differences in words exchanged and notes written could result in changes in healthcare providers' actions and patient reactions. The success of the next version of the prototype being more clinically effective than the previous one depends on the accuracy with which inventors can anticipate and adjust vis-à-vis human responses to the device much more than their ability to tweak the material components of the hardware. And this lack of cultural knowledge of the clinical contexts, patient needs, and doctor expectations is what initiates many Indian engineers' and designers' desire to improve or replace imported medical devices with indigenous innovations.

While prototypes suggest an entity—object, method, or idea—that will emerge in the future, many devices are forever stuck in such early stages of creation. Inventors often thoughtfully design efficacious devices that achieve desired clinical outcomes, yet their prototypes may remain off-market if they do not find interested corporate buyers and manufacturers, or are unable to obtain government procurement for public hospital distribution. Medtech prototypes that do not become scaled up into mass-produced, widely circulated health technologies often reside in the garages, attics and shelves of entrepreneurial spaces. Discarded prototypes are ghostly objects, memorialized within the startup team's collective memories, but unknown to anyone outside of their social circles.

The small social and material scales of prototype production make it distinct to many studies on contemporary South Asian biocapital. A rich anthropological literature—on pharmaceuticals, surrogacy, organ donation, and other issues—has revealed the way biocapital rooted in South Asia is produced, commoditized, and transacted in social life and on the global stage (see Bharadwaj and Glasner 2012, Rudrappa 2015, Sunder Rajan 2009, Venkat 2016). This research has largely focused on how the Indian biomedical technology industry is marked by generic manufacturing and the commodifed exchange of mass-produced biological products. In the case of health startups, however, similar bioethical questions arise, yet in the entirely different context of innovation—where value stems from novelty and unorthodoxy. Many of my interlocutors bemoaned Indian higher education, especially state-run public schools, for emphasizing rote memorization over creative thinking. Inventors thus have great incentive in keeping unscaled prototypes, even though there might only be a handful of physical copies. They might own patents, and often, the barrier is in proving the device will be its marketability in later stages of the invention process, not in its efficacy, which has already been verified during the final stages of prototyping.[4]

Discarded prototypes, or ones that are temporarily put aside in hopes that they will be marketed in the future, require materials and interactions that still reveal bioinformation, only through social relations of producing, rather than using, a technology. Early-stage prototyping of novel medical devices produces the antithesis of big data volumes. Yet even when only a small number of a prototype are manufactured, hardware materials are constructed through social relationships among and between inventing teams and their test subjects, resulting in bioinformatic data feedback between people and devices. Small-scale production of unique material objects requires practices, ethics and logics that I argue are just as reflective of bioinformation as studies of big data. As I will show in the following section, a handful of prototypes may still come in contact with myriad people's bodies, further propagating bioinformation feedback and circulation vis-à-vis tech production.

Bioinformation, in formation

The current Indian administration has fostered the medtech sector through allocating resources for medical device making in public higher educational institutions and hospitals, as well as growing a startup ecosystem to provides funding, materials and workspace to many Indian entrepreneurs (Goswami 2017, Johnson 2016). The Indian Institutes of Technology (IIT), founded across India at the time of independence, were initially envisioned as institutions that would train engineers to serve the new nation's technological development agenda (Subramanian 2019). When they took power in 2014, the current BJP administration launched the *Make In India* and *StartUp India* campaigns, widely advertised as "nation-building initiatives designed to transform India into a global design and manufacturing hub."[5] As Banu Subramaniam shows, the Modi administration has "embraced capitalism and Western science and technology as elements of a modern Hindu nation" (2019: 7), and this government's onus for technological innovation and social impact by citizen scientists has since increasingly been placed within the entrepreneurial sector (Irani 2019).

Through their dependence on government funding, and laboratory and hacker space access, Indian medtech inventors continue to entrench such nationalist development projects. Healthcare startups in particular depend extensively on public-private partnerships, as they additionally utilize public hospitals to find human test subjects. Public institutions thus become the spaces where many Indian inventors move through initial stages of ideation, prototyping and testing. Similar to new translational medicine research initiatives in other nations, the early stages of Indian medtech experimentation and trials routinely occur in public institutions and through public funds, yet the resulting products are often sold in commercial markets (see Robinson 2019). In the

Indian context, many devices are designed for the particular resource and infrastructural constraints found in the public health system, yet entrepreneurs may obtain other clients at the time of commercialization. Before these late-stage departures, however, prototyping nearly always occurs in connection to public resources and institutions. And, as we will see in the following ethnographic example, entrepreneurs usually depend on patients who access these systems for their healthcare.

As I climbed to the second floor converted Bangalore bungalow that served as a medtech company's headquarters, I was struck by how many of the startups I had now visited worked out of rented residential spaces. The house was in a row of square bungalows very similar to my own housing two kilometers away, shaded by old leafy fig trees, with a terracotta roof, narrow stairs that led above the ground floor car parking to the main house, and dusty pastel pink paint. An electrical engineer and industrial designer turned inventor, Raj, had invited me to the space to see his neonatal hearing device firsthand, and meet his team: a few engineers, a doctor and an audiologist.[6] We began talking and Raj said:

> It's good to know you're more interested in the genesis part of medical device innovation … In a country like India where you have a scarcity of resources, you should invest more time in preventing problems rather than treating problems.

He continued, telling me how he initially became interested in inventing in the ENT space after participating in a government-sponsored health innovation program, where he observed hearing loss in children who were brought to a clinic in which he was assigned to do observations.

> A parent walked in with their five-year-old son and I was completely unaware of this problem at the time, and they just mentioned that the child was not able to speak. The doctor, after examining, said he's not able to speak because he's not able to hear. And it's already too late to do any intervention, so you can just go and get the disability certificate done … I can kind of relate to that, because it happens so often in different parts of India … In the first two years [of life] it's crucial that hear the environment, that's how our mind learns to speak … if you have to do an intervention, you should do it in the first six months. That's why in advanced systems, like the US or Europe, they do the [hearing] screening of each and every baby at the time of birth. So that's the process all over the world, but it's not happening in India or developing countries … and the majority of babies born with hearing loss are in the developing world.
>
> (Interview, 24 May 2019)

Raj went on to explain that these observations led him to want to meet this clinical need in India, specifically designing his device for resource-poor clinical sites around the country. After our conversation, he showed me around the office. Like many similar spaces, the kitchen was easily converted into a wet lab, the bedrooms into brainstorming nooks or spaces for engineers to tweak the sensors and plastics of the devices. The only room in which he requested I not take notes was a hardware experimental space where he told me a new idea for the prototype was in progress. At the end of the day, Raj extended an invitation for me to accompany the startup's audiologist, Varsha, to the upcoming weekly device testing sessions, where a handful of the early-stage devices were being tested in a nearby hospital.

One morning the following week, I thus found myself outside a large government hospital waiting for Varsha to arrive. My *kurta* and *chooridar* stuck to my body in the humid Bangalore summer heat. I moved under the shade of one of the awnings that lined the sides of the large government hospital. Patients hurried in and out of the ward, accompanied by visibly worried family members. As with all public medical institutions in India, this one offered free or highly subsidized care, attracting patients who cannot afford private hospitals. It bears mentioning that the majority of people in India depend on the public health system's care, which has a complex history of accessibility and efficacy (see Kothari and Metha 1988, Kumar 2010, Ludden 2005). Many patients and families from across the state of Karnataka, and often farther, traveled to this particular hospital every day. Varsha soon pulled up on her *scooty*, nodding to me to hop onto the backseat. We zipped to a nearby ward where we would spend the day testing the prototype with patients.

I had seen Varsha in medtech events across the city before we crossed paths again in the bungalow the previous week. She and I got along easily, being of similar age and sharing the experience of growing up in Indian immigrant families in the US. After almost a year of fieldwork, it struck me that this was the first time I would be spending an entire day working with a fellow woman. My objective in attending the weekly hospital session was to observe the ways the prototype might be refined between tests; Varsha seemed quite happy at having company and the extra set of hands. The team had been working on this particular device for the several years, and like all medtech inventors, they are regularly checking a few units against human subjects. Raj and his team were at the late stages of inventing this particular device; they had recently received ISO approval through the Indian government, and were waiting to receive FDA approvals or CE mark. Ensuring efficacy in providing outcomes that are satisfactory to doctors, gauging prospective patients' reactions to their novel technology, determining its accuracy in producing replicable and verifiable data, and discovering any unforeseen issues in administering the device in similar clinical settings, were of key importance before the device could be distributed even more widely.

"Look like you belong, walk like a doctor," Varsha instructs as she ushers me past a security guard and into the building. I do not have time to comprehend how to do this as I hurriedly follow her up a flight of stairs and into a sea of women, each holding a newborn baby. We are in the neonatal ward, she explains, as we cut through the crowd to a set of doors on the far side of the corridor. Today, we will be testing the hearing of premature infants.

Simultaneously, we will be testing the efficacy of the device prototype in its current form. Entering the testing room, Varsha instructs me to bolt the door behind us. The room is furnished sparsely: two steel framed beds, with plain mattresses covered in old, yet clean, thick cotton covers; a few plastic chairs; a baby weighing scale that is missing the basin; a set of gray padlocked lockers; and a toilet and sink, separated from the rest by a temporary wall partition.

Varsha sets up on one of the beds, carefully laying out the device, a dense plastic square with a small screen and soft buttons, proceeding to connect the electrodes that will in turn connect the babies to the device. She lays out cotton, notebooks, gel to clean the electrodes between infants, taking care to leave a small area in the middle of the bed to place them. A doctor affiliated with the project, whom I assume is an obstetrician, enters the room. Varsha introduces us, as the doctor arranges three large stacks of patient files on a small table and hurriedly leaves. We hear the chatter of the mothers, all anxious for their babies to have their hearing tested, each time she opens the door to enter and exit. The doctor leaves the patients in our care—or our "intervention" as the entrepreneurs call it—saying she will check back in later.

Varsha and I proceed to go through the files methodically, attempting to keep the testing room quiet by only letting in two patients at a time. But within an hour we give up on this system. Every time we open the door to call another name, women come in and out haphazardly, and soon the room is brimming with mothers and crying infants. I watch each mother place her baby in the middle of the bed—these are premature infants, two or three pounds each, all swaddled heavily in blankets. We gingerly unwrap them while Varsha delicately places the electrodes on each baby's head and calibrates the hardware that takes the input from the electrodes and produces the readings on its small screen. As the sound level in the ward is fluctuating uncontrollably, I am not sure if the data outputs are the babies' brains registering the specific sounds we are sending through a headseat, or the persistent din from outside the testing room. Varsha corroborates what Raj had told me the week prior, that the device is designed for hospital environments like this, where a soundproof room is inaccessible. However, as she bemoans the noise level in the room throughout the morning, I do my best to direct mothers in and out. To complicate matters, Varsha explains that the infants born with hair are registering sound interference, something we note could be improved if the device is re-designed.

I notice a series of post-it notes on the wall next to the bed where someone has detailed the device testing protocol. Varsha keeps encouraging me to participate, and I hesitate. I am not clinically trained, I keep reminding her, very nervous about doing any wrong by a premature infant. I settle into the role of helping keep the babies calm, gently holding them while she arranges the sensors across their foreheads and notes the neurological output. Additionally, as I was the only Hindi speaker working with Varsha that day, I was glad to be able to be of service communicating with the mothers who only spoke Hindi. Varsha called out readings, and I helped her make pencil scratches in the patient charts. The startup team was trying to make the device as comfortable, quick-registering and accurate as possible. Raj had told me that one of the reasons he chose to invent for this particular clinical need was because the device was non-invasive, which made it relatively safe and low risk.

Varsha and I note the patient data, and also the ways we can tweak the device's sensors and cross-check the accuracy of the readings, for device improvement. The device relies on a data collection technique, termed Brainstem Auditory Evoked Response (BAER), which has become the standard hearing measurement metric in Euro-American pediatrics practice since the 1970s (Jewett et al. 1970). The apparatus it requires, however, is prohibitively expensive in many Indian health settings. The team's aim was to tweak the prototype to match standardized BAER data outputs, and have the finalized device used by hospitals in resource constrained settings, whether through public or corporate acquisition. Testing the prototype in hospital with Varsha certainly illuminated moments where the BAER readings were not what she hoped. Yet it also illuminated myriad bioinformation feedback, such as communicating with the mothers, the soundscape of the testing room, and variations in infants' physical features, that inform how the design could be altered.

In this way, bioinformation needs to be understood as it is produced in relation to medical, engineering and design practices in the Global South. Through these practices, a form of bioinformation specific to cultural logics and biomedical ethics of Indian medtech is taking shape. Once perfected, medtech prototypes may be marketed in their final forms as low-cost responses to universal health dilemmas. Yet their very formation is predicated on social relations and political-economic histories specific to South Asia. As we see in this example, these relations and histories can reverberate with longstanding exploitation of often resource-poor, low-caste, patients turned experimental subjects. Therefore, we can recognize bioinformation as the data technologies garner once they circulate in (sometimes global) markets. But we also must recognize bioinformation in the bodily outputs that provide the conditions for the production of new devices in the first place, often erased from the development narrative by the time devices are commercialized.

Producing objects, producing subjects

The example of testing the neonatal hearing device with Varsha reveals prototype testing as inherently rife with ethical questions about patient safety, privacy and the collection and storage of data. In particular, the production of technologies yields bioinformation by turning human bodies into experimental test subjects. Yet, paradoxically, without this bioinformation, devices required for effective patient care in India would not be eventually deployed into resource-poor clinical settings where patients need them most. Further, as this bioinformation is constantly analyzed by inventors and fed back into their lengthy tech production practices, it turns humans into entrepreneurial subjects as well. Thus, this circular relationship between technology, data, and human bodies is amplified as entrepreneurs progress further into the process of making a prototype.

The number of times a prototype is brought back to be tested in a public hospital takes on additional salience in the Indian context, as prototyping often means elite, male, high-caste techies test their objects on often more marginalized patient populations in the government public health system, overwhelmingly women looking to access health services for themselves and their children. While these stark power differentials regarding gender, class and caste are not to be ignored, I hesitate to portray prototype testing in a purely negative light. Medtech prototyping often results in efficacious outcomes for test subjects. In my experience with the neonatal device startup, many mothers would not have accessed any hearing screening for their babies if it were not for the device testing arrangement between the public hospital and the startup, due to resource constraints. Potential risk, harm, and benefit to patients is rarely black and white. While their ethical regulations may not coincide within the Euro-American frameworks of informed subjects protocol, Indian medtech inventors and their overseeing clinical partners weigh similar ethical factors alongside their deep cultural understandings of the particular settings in which they work.

Science studies scholar Lilly Irani demonstrates how logics of what she calls "design as development" are being used in contemporary urban India to shift the responsibility of social change and upward mobility away from social services and instead towards private citizens as problem-solving innovators (2019). Instead of investing directly in public healthcare, the current Indian administration is investing heavily in the entrepreneurial subset of the private sector—one that is marked by global flows of capital, people, and intellectual property—to address public health concerns. Indian entrepreneurs are thus working at the intersection of the purported increase in Indian public health services and the increasing neoliberal privatization represented by the global startup tech industry. In this shifting context, governmentality is mediated through new public-private assemblages and mobilized through a heterogenous mix of both nationalist development discourses and universal humanitarian discourses, that are not necessarily at odds with each other.

Scholars of South Asia have demonstrated the social and environmental porosity and mutability of the human body (Solomon 2016). However, the value of new medical devices—"unicorns," as unique, new ideas are often referred to the startup world—requires assuming test subjects have uniform physical compositions. Engineers and doctors trained in scientific paradigms historically depend on controlled experiments, yet they do not always adhere to meticulous or organized record-keeping at the earliest stages of testing prototypes. Instead, they often assume that testing a slightly tweaked prototype on a different person or population will produce results replicable *enough* to move forward in the invention process. Varsha, for instance, continued testing infants even after we established that issues with the room might affect data outputs. This form of testing simply collects data from bodies to justify progressing to a subsequent step in device creation, or spending more time solving a particular problem with the current model. However, within these trial-and-error proceedings, entrepreneurs often produce numeric biometric data that concerns doctors and patients and precipitates medical intervention. In these cases, for example, if a baby was found to have problems hearing, the mother would be told so she has the option of seeking out treatment before leaving the hospital.

Gender, caste and class differences deeply shape human participation in the Indian medtech community, not only among entrepreneurs, but through their relationships to their experimental subjects. Individual consent and institutional oversight operate through unique ethical frameworks in India. Indian entrepreneurs' knowledge of the specific bureaucratic and infrastructural hurdles enables them to tweak devices-in-progress for successful uptake in clinical settings where imported devices were unusable because they were inaccessible, prohibitively expensive, or infrastructurally incompatible. Yet it also allows medtech entrepreneurs to negotiate testing devices within familiar health settings and effectively navigate the historically impenetrable Indian public healthcare system. Interestingly, many Indian medtech entrepreneurs claim that they decided to invent in and for India for such logistical reasons. They described unenforced regulations, as well as the sheer numbers of public hospital patients, that contributed to a productive pace of device innovation, and made the initial capital investment or grant procurement in India feel less risky than other places. Of course, the counterfoil to this solution might be decreased risk management and oversight for clinical trial participants. The practice of prototyping medtech devices thus yield a mix of humanitarian and capitalist logics, all too common in medical intervention (Fassin 2011, Redfield 2013).

Biotech companies have depended on global experimental test subjects for decades (Petryna 2009), often assuming the Global South can serve as a petri dish, an experimental landscape before therapeutics are produced for Euro-American consumers (McNeil 2005, Roy 2017, Tilly 2011). However, in recent years, local efforts that lay claims to intellectual property over Indian medicines have subverted some multinational pharmaceutical companies' interests (Halliburton 2017). While small Indian medtech companies often depend on similar experimental test subjects as does big pharma, medical device producers often invoke nationalist discourses, rendering themselves in closer relation to those on whom they test their prototyped technologies. The smaller scale of medtech versus biotech production, especially during prototyping, sheds light on entrepreneurs' assumptions that interchangeable bodies serve as experimental constants on which the effectiveness of the same prototype can be tested. Yet the heterogenous and often chaotic circumstances in public hospitals also reveal another imperative of medtech entrepreneurs' voice: their resolve to hack particular Indian health issues with cultural knowledge of infrastructural constraints and indigenous design.

Experimental data produced through prototype testing has built-in hazards. Even as it is monumental in determining which medical devices are eventually produced by the global tech industry, un-standardized and sometimes chaotic methods of recording and collecting may limit its longer-term use. Data garnered from bodies during the early stages of device invention sometimes feeds into the complex of big data, and sometimes does not. "Small" amounts of un-replicable, unrecorded data can also have the potential to travel far and live on within various infrastructures. If the invention process is successful, these nuanced connections between body and data will be amplified to create finalized medical devices, biocapital that travels to a variety of clinical users in India and beyond (Vora 2015). Trial-and-error prototype experimentation not only drives this collection of bioinformation that has the potential to be scaled up, but is also a necessary driver of medtech experimental progress. Ethnographically tracing the exchanges between inventors and the bodies on which they choose to test their technologies in-formation thus reveals how bioinformation is produced, but also produces new technologies and new subjects.

Conclusion

As states, NGOs and tech industry companies—corporations and small startups—are monitoring and analyzing consumer data, it is imperative for anthropologists to focus on the social relations through which data is garnered from bodies (Parry and Greenhough 2018). The health niche of the global tech industry provides a generative context to study bioinformation ethnographically. Tracing the minutia of tech hacking illuminates how bioinformation travels in smaller circuits of interpersonal, visceral engagements, before it gains the potential to be scaled up into big data sets.

The prototype stage of the invention process demonstrates the ways small bits of bodily data can be amplified through the tech production process, which is necessary for both new devices and new patient, doctor and inventor subjectivities to take shape. However, even if the initial stages of the invention process do not yield a successfully marketed technology or go on to proper clinical trials, such instances of bioinformation are still rife with ethical questions. I showed in this chapter how anthropologists can engage with the fraught semiotic-materialities of new forms of health data, as these technologies circulate in India and globally. Paying special attention to the diverse Global South contexts where urgent issues of consent, regulation and intervention politics vary, I have highlighted how myriad social relations between contemporary Indian subjects and emergent technologies undergird the production of medical bioinformation and endow it with local and global meanings.

Inventing new medical devices in India requires both a production and feedback of bioinformation in order for devices to become finalized, efficacious, marketable technologies. Entrepreneurs extract data from experimental subjects, and from themselves, but all to be fed back into a loop that allows for the technology be formed. This results in devices, experimental subjects, and entrepreneurial producers being co-produced alongside one another (see Prentice 2013). Overall, bioinformation gathered, manipulated and produced in Indian medtech innovation thus demonstrates how technologies are always co-produced with complicated forms of data, and new types of people.

In this chapter, I have illuminated how the production of novel medical devices occurs through the testing of prototypes on people at each stage, showing how a collection and production of bioinformation is central to the ways entrepreneurs envision themselves progressing through subsequent technology invention stages. Medtech inventors require interactions with human subjects to understand if a device in progress is becoming what was originally conceptualized, and this imagination cannot be realized by working on the material prototype alone. Straddling moral imperatives to provide accessible health screenings yet also capitalist motivations to produce tech quickly and affordably, Indian medtech creation produces bioinformation throughout the device creation process. Even more notably, it depends on bioinformation derived from real-time humans eager to access valuable resources in order to come into the world.

Ethnographically examining medtech prototyping in India reveals permutations of relationships between different actors—entrepreneurs, test patients, raw hardware materials, prototypes, finalized devices, the state—that together constitute bioinformation. Some of these relationships are transient, others enduring, but all contribute to iterative cycles of bodily co-presence producing new data, new technologies, and new Indian subjects. The virtue of anthropological fieldwork in the Indian medtech community is to shed light on the social relations that are crucial to invention, yet often erased from high-tech development narratives that ignore where data originates once final products are globally marketed. This under-theorized role of human relations is continuously best revealed in granular investigations that attend to the semiotic as well as material dimensions of prototyping.

Notes

1 "Drawer walls," as I came to refer to them, are a recognizable facet of India's medical device sector.
2 Hereafter, Indian Institutes of Technology will be referred to by their widely used abbreviation: IITs.
3 For example, across the medtech ecosystem, entrepreneurs excitedly recounted the legendary story of Tom Fogarty, who invented the first balloon catheter through casual observations of surgical practice while he was a scrub technician. The resultant device was completed before he graduated medical school, and Fogarty ascribes his success to his understanding of basic hardware components and fishing expertise, like tying knots (Brown 2002).
4 This is often obtained before the device is created, as entrepreneurs want to have a legal claim on their ideas. Many patents do not actually become material devices, but the patents can still be enforceable for years, if not decades.
5 As cited on the Department for the Promotion of Industry and Internal Trade website, the forum of the Indian Central Government's online presence that manages these campaigns. Updated on and accessed on 1 September 2020.
6 All names have been changed to protect privacy.

References

Bharadwaj, A. and P. Glasner. 2012. *Local Cells, Global Science*. New York: Routledge.
Bhatt, Amy. 2018. *High Tech Housewives: Housewives, IT Workers and Transmigration*. Washington: University of Washington Press.

Brown, David E. 2002. *Inventing Modern America: From the Microwave to the Mouse.* Cambridge, MA: MIT Press.

Chaturvedi, Jagdish. 2015*Inventing Medical Devices: A Perspective From India.* CreateSpace Publishing.

English-Lueck, J. A. 2002. *Cultures@SiliconValley.* Palo Alto: Stanford University Press.

Fassin, D. 2011. *Humanitarian Reason: A Moral History of the Present.* Berkeley, CA: University of California Press.

Goswami, S. 2017. "These 3 Healthcare StartUps are Tapping Into India's Soon-To-Be Booming IoT Market." *Forbes*, February 22.

Goyal, Malini. 2019. "How IT Graduates Are Giving Up Their Dollar Dreams For Startups and New Age Companies." *Economic Times*, December 15.

Halliburton, Murphy. 2017. *India and the Patent Wars: Pharmaceuticals in the New Intellectual Property Regime.* Ithaca, NY: Cornell University Press.

Haraway, Donna. 1988. "Situated Knowledges: The Science Question in Feminism and the Privilege of Partial Perspective." *Feminist Studies* 14(3): 575. https://doi.org/10.2307/3178066.

Irani, Lilly. 2019. *Chasing Innovation: Making Entrepreneurial Citizens in Modern India.* Princeton, NJ: Princeton University Press.

Jewett, D. L., M. N. Romano and J. S. Williston. 1970. 'Human Auditory Evolced Potentials; Possible Brainstem Components Detected on the Scalp'. *Science* 167(3924): 1517–18. https://doi.org/10.1126/science.167.3924.1517.

Johnson, T. A. 2016. "Bengaluru Sees A Wave of Health Tech Start-Ups As Innovators Turn Life Savers." *The Indian Express*, August 16.

Kirschenbaum, Matthew. 2008. *Mechanisms: New Media and the Forensic Imagination.* Cambridge, MA: MIT Press.

Kothari, Manu L. and L. A. Metha. 1988. "Violence in Modern Medicine." In *Science, Hegemony and Violence.* A. Nandy, ed. Oxford: Oxford University Press.

Kumar, Deepak. 2010. "Probing History of Medicine and Public Health in India." *Indian Historical Review* 37 (2).

Latour, Bruno. 1988. *Science in Action: How to Follow Scientists and Engineers Through. Society.* Cambridge, MA: Harvard University Press.

Ludden, D. 2005. 'Regimes in South Asia: History and the Governance Conundrum'. *Economic and Political Weekly* 40(37): 4042–51.

Marcus, George. 2014. 'Prototyping and Contemporary Anthropological Experiments With Ethnographic Method'. *Journal of Cultural Economy* 7(4): 399–410. https://doi.org/10.1080/17530350.2013.858061.

McNeil, M. 2005. "Introduction: Postcolonial Technoscience." *Science as Culture* 14 (2).

Murphy, Keith. 2016. "Design and Anthropology." *Annual Review of Anthropology* 45.

Ogrodnik, P. 2012. *Medical Device Design: Innovation From Concept to Market.* Cambridge, MA: Academic Press.

Parry, B. and B. Greenhough. 2018. *Bioinformation.* Cambridge, MA: Cambridge University Press.

Petryna, Adriana. 2009. *When Experiments Travel: Clinical Trials and the Global Search for Human Subjects.* Princeton, NJ: Princeton University Press.

Prentice, Rachel. 2013. *Bodies In Formation: An Ethnography of Anatomy and Surgery Education.* Durham, NC: Duke University Press.

Redfield, Peter. 2013. *Life in Crisis: The Ethical Journal of Doctors Without Borders.* Berkeley, CA: University of California Press.

Roy, R. D. 2017. *Malarial Subjects: Empire, Medicine and Nonhumans in British India 1820-1909.* Cambridge: Cambridge University Press.

Rudrappa, S. 2015. *Discounted Life*. New York: New York University Press.

Shankar, S. 2008. *Desi Land: Teen Culture, Class and Success in Silicon Valley*. Durham, NC: Duke University Press.

Solomon, Harris. 2016. *Metabolic Living: Food, Fat and the Absorption of Illness in India*. Durham, NC: Duke University Press.

Subramaniam, Banu. 2019. *Holy Science: The Biopolitics of Hindu Nationalism*. Washington: University of Washington Press.

Subramanian, Ajantha. 2019. *The Caste of Merit: Engineering Education in India*. Cambridge, MA: Harvard University Press.

Sunder Rajan, Kaushik. 2009. *Biocapital: The Constituion of Postgenomic Life*. Durham, NC: Duke University Press.

Tilly, H. 2011. *Africa as a Living Laboratory*. Chicago, IL: University of Chicago Press.

Venkat, B. 2016. "Cures." *Public Culture* 28 (3).

Vora, Kalindi. 2015. *Life Support: Biocapital and the New History of Outsourced Labor*. Minneapolis, MN: University of Minnesota Press.

Woese, Carl. 1987. "Bacterial evolution." *Microbiology Reviews* 51 (2).

Yock, C. *et al*. 2015. *Biodesign: The Process of Innovating Medical Technologies*. Cambridge: Cambridge University Press.

6 Top_to_toe.ods

Bioinformation and the politics of rape response

Sylvia McKelvie

Rape belongs to a crowd and it also belongs to the household. Rape is for groups and for individuals. Rape is historic and daily. Rape is for the ugly as well as the beautiful. Rape does not belong to one people or only in the time of war or only in the time of peace it is not the crime of strangers and it is not only the crime of statelessness it is also the crime of the state. Rape is not only anarchy it is also a social order. Rape is of nature but also not of it.

(Anne Boyer, *My Common Heart*)

On 1 November 2018, the direct-action group Sisters Uncut blocked the doors of the Crown Prosecution Service (CPS) in London. Protesters piled approximately 30,000 sheets of paper bound in legal tape and stamped with "CONFIDENTIAL". The pages represented the average amount of personal data that survivors of sexual assault and abuse were required to submit during criminal investigations. Known as Stafford statements in the UK, the police and prosecution issued these requests in order to access victims' data. Digital data included social media, web activity, instant and SMS messaging, location data, emails, deleted data, images, videos, audio files, apps, contacts and documents. Additional personal documents included counselling notes, reports from social services and educational records.

In July 2020, it was announced that digital data "consent forms" would no longer be used during sexual assault and rape cases (Topping 2020). Yet the legacy of Stafford statements and the seizure of personal information is one of many practices that make up the data worlds of sexual violence. Understandings of data and sexual violence are largely associated with the criminal justice system. However, over the last several years increasing attention is being paid to sexual assault and abuse as public health concerns. A report from the US Centers for Disease Control and Prevention (CDC) outlines a comprehensive list of data elements that constitute effective record-based and survey surveillance of sexual violence, covering victim demographics, perpetrator details, assault-related findings and post-assault outcomes such as mental health and medical care. The CDC's (2014) recommendations echo the belief that the more we know, the greater the collective ability to prevent sexual violence.

DOI: 10.4324/9780367810030-6

Today, public concern is attuned to the management of bioinformation and personal data in sexual assault cases. The rape kit backlog has made international headlines and prompted outcry from activists and survivors. National campaigns and programmes in the US, such as the Sexual Assault Kit Initiative (SAKI), have been developed to test hundreds of thousands of kits and prevent future backlogs. In a comparable response, the House of Lords called for major improvements in the provision of forensic sciences in the UK. An article in the Guardian from May 2019 declared a "crisis in forensic science," with labs "on the brink of collapse," suggesting that better funding for and regulation of forensic labs will minimise delays for processing bioinformation and digital data (Devlin 2019). In both instances, it is argued that streamlining governance and increasing capacity will lead to more convictions. The backlog and capacity crisis are framed as bureaucratic problems, implying that if campaigns and governments can improve management then there will be justice for survivors.

The handling of forensic evidence by different actors and authorities articulates what it means to endure, disclose and report sexual violence. This chapter looks at converging narratives of bodies-as-evidence and bodies-as-information, interrogating the transmission of medico-legal "truth" through bioinformation. Drawing on sociological and anthropological scholarship, I focus on new jurisdictions and frontiers of data collection, use, analysis and retention, and the fragmentation of sexual trauma across backlogs, biorepositories, and databases. How do these material and cultural practices figure victimhood? Further, how does bioinformation transmute *what* rape is?

In this chapter, I also articulate a feminist analysis of bioinformation. As history shows us, mainstream anti-rape activism was integral in the design and delivery of state-led responses to sexual assault, including standardising forensic tools and knowledge (Bumiller 2008, Quinlan 2019, Shelby 2018). As Wang suggests, data like bioinformation "installs itself as a solution to the problem of uncertainty by claiming to achieve total awareness," rising out of our human-bound limitations (2020: 238). Technoscientific developments that collate, classify and capture sexual violence seem to "modernise" justice—and yet, to paraphrase Benjamin (2019), in many ways these advancements are forms of social control. The making of the capacity crisis, which I would argue is also a crisis of carceral feminism, brings forth new questions for feminist technoscience and bioinformation studies.

State bindings: epistemic challenges of the anti-rape movement

Picturing a forensic laboratory, one imagines a sea of white coats and microscopes. If you add the additional filter of sexual assault, then you're flooded with media representations of stranger rape and miraculous DNA matches. Bioinformation is of course a part of this visual narrative, distilled as biological trace left on the victim's body. As Steenberg (2013) states, this body is usually female, murdered and sexually violated. Culturally, we accept these representations as fact. The epistemological arguments of forensic science have largely avoided the same criticism as other disciplines. Steenberg writes, "Forensic science, whether on reality television or in a video game, remains fixed in a period of Enlightenment humanism and frequently elides the controversies, anxieties and inconsistencies circulating around science in postmodernity" (2013: 16).

Contemporary public anxieties around backlogs and laboratory efficiency fixate on the supremacy of forensic science (Quinlan 2019). Like unclogging an artery, it is suggested that with reform the system will function as intended. State-led initiatives like SAKI also garner broad support from feminist and anti-violence organisations (Quinlan 2019). At first glance, this support might seem in the best interest of survivors. But what is often absent from current conversations around forensic provisions is an understanding of the complex history of the "feminist alliance with the state," which includes forensic authority and police power (Bumiller 2008: 2). Before I delve into the meanings given to bioinformation collection, use and retention, I want to illuminate the history of "rape response" as well as the tensions within scholarship on this very topic—how sociologists and anthropologists think through technoscience and the politics of gender-based violence?

As Corrigan outlines, the anti-rape movement took to the legal battleground in the 1970s and 80s, fighting for substantial reforms to criminal law. Though this history varies country to country, activists focused on improving their respective legal systems so that victims were not only heard in the courtroom, but that rape was seen as a social problem (Corrigan 2013). Anti-rape groups challenged sexual psychopath statutes and argued for less draconian sentences, hoping that more lenient thresholds would increase convictions and protect future victims (Corrigan 2013). With the rise of DNA testing, victims' groups became interested in forensic evidence as the "currency of the court" (Shelby 2018: 8). Protocols were developed for collecting clothing, inspecting victims' bodies and streamlining the overall handling of evidence (Shelby 2018). As early as 1975, a prototype rape kit (also known as a sexual assault examination kit) was developed in the US (Shelby 2018).

Scholarship shows that feminists have had a somewhat uneasy relationship with the influence of forensic science. For example, in the 1990s feminists in Canada began questioning how forensic evidence was used during rape trials, citing some troubling trends (Quinlan, Fogel and Quinlan 2010). DNA evidence was preferred over the testimony of victims and even when it was available, defences would adopt consent arguments, suggesting that forensics only prove that intercourse took place, not that it was non-consensual (Quinlan, Fogel and Quinlan 2010). Nonetheless, these evidence collection tools would become commonplace in hospitals in Canada, the US and the UK, signalling a closer alignment of forensic, medical and legal authorities (Shelby 2018).

The early history of rape response is also about the institutionalisation of feminist approaches to sexual violence and health (Corrigan 2013, Caringella 2009, Murphy 2012, Shelby 2018, Quinlan 2019, Gruber 2007). Technoscientific interventions like the rape kit brought "rape work" into the medical arena (Shelby 2018, Martin 2005: 9). If tools like rape kits were to be implemented, anti-rape and feminist organisations, such as rape crisis centres and grassroots health clinics, recognised the need for specialist training and support. The alliance that Bumiller (2008) describes laid the groundwork for programmes like Sexual Assault Nurse Examiners (SANEs), Sexual Assault Response Teams (SARTs) and Sexual Assault Referral Centres (SARCs). These programmes were designed to provide acute care, which regularly involved offering forensic medical examinations (FMEs). Anti-rape advocates and crisis workers would accompany victims, supporting them during the physical examination and facilitating their engagement in the criminal justice system, which is still common practice today (Martin 2005).

While the delivery of sexual violence support varies state-to-state and even municipality-to-municipality, the literature shows us that as feminist approaches to sexual violence responses became institutionalised, in turn, the responses to sexual violence were depoliticised. Baker and Bevacqua characterise this turn in feminist inquiry as a "narrative of decline" (2018: 351). In reference to the work of Rose Corrigan, Kristin Bumiller and Aya Gruber, Baker and Bevacqua argue that "decline" literature suggests that medico-legal reforms and cooperation with social services neutralised rape work, making greater accommodations for state involvement and priorities.

What scholars like Rose Corrigan and Kristin Bumiller demonstrate are the ways in which stereotypes about sexual assault are reinforced across jurisdictions and in everyday processes; medical, legal or otherwise. Decline (or depoliticisation) literature argues that even with substantial reforms to legislation and the introduction of sexual violence specialisms, survivors consistently experience re-traumatisation by medical, legal and other actors. Further, their work also rewrites the history of feminist advocacy, either by exploring its neoliberal co-option (Bumiller 2008), its decline in anti-rape advocacy (Corrigan 2013) or its support for mass incarceration agendas (Gruber 2007, Gruber 2020).

In contrast to arguments of decline, Egan (2019) suggests that feminist politics have been so effectively woven into the sexual assault sector that they're commonplace. Referencing guidelines set out by Australian health authorities, Egan (2019) highlights the mainstreaming of sexual assault definitions, which emphasise coercion, power and the danger of victim-blaming attitudes. However, Egan's (2019) analysis fails to consider the intersection of oppressions; *whose* feminism is privileged in these spaces and services? Phipps (2020) response would be bourgeois white feminism, arguing that movements like #MeToo have been co-opted, focusing on protection for survivors of a certain class, race or sexuality.

Like Phipps, Baker and Bevacqua (2018) present another reading of the anti-rape movement and its history; rather than being in decline, there is a burgeoning movement that aims to fight systemic and institutional forms of violence at the intersection of gendered *and* racialised abuse. Community-based activism and interventions that challenge the mainstream women's movement—and its carceral logics—are grounded in the foundational work of Black abolitionist feminists (Baker and Bevacqua 2018). In a recent anthology edited by INCITE!, Rojas Durazo writes that anti-violence groups working at the "intersection of state and interpersonal violence" regularly face resistance and even attacks within the non-profit industry (2017: 122). Rojas Durazo's (2017) arguments, much like Phipps (2020), suggest that only certain feminist standpoints have been subsumed by mainstream rape response. Echoing the writings of Crenshaw (1991), Rojas Durazo (2017) problematises sexual and domestic violence law reform and its inability to speak to lived experiences of racism, sexism and classism.

While one can't disregard the contributions of Rose Corrigan, Kristin Bumiller and Aya Gruber, Baker and Bevacqua (2018) identify a tendency in studies on sexual violence to stumble into a phenomenological deadlock, or the practice of emphasising and *re*-emphasising violence that is inherent within cultural and institutional discourses. This chapter may fall into a similar habit because by its very nature, sexual assault is a minefield. As Mithu Sanyal writes in *Rape: From Lucretia to #Metoo*, "Rape is a veritable hall of mirrors of expectations and discourses, and each sentence is followed by ten unspoken ones. I call this a cultural sore spot. Like sore spots on the body, cultural sore spots indicate something that needs our attention but that we are afraid to touch" (2019: 2).

Scholarship is not afraid to touch per se, but sexual violence continues to be a polarising issue and rape itself an "'essentially contested category' infused through and through with political meaning" (Bourke 2015). Critical inquiry into contemporary responses to sexual violence is given a moral responsibility to confront the compendium of rape myths and must be careful as well to not undermine the contributions of rape work and legacies of feminist organising (Martin 2005). Much like decline literature, this chapter is interested in the depoliticisation of rape response and to this end, I position bioinformation and sexual violence data as underexamined fields of power. Further to the work of Baker and Bevacqua (2018), Rojas Durazo (2017) and Phipps (2020), this chapter looks from technoscientific encounters to larger regimes of oppression and repression. As postcolonial scholar Sherene Razack contends, "when the terrain is sexual violence, racism and sexism intersect in particularly nasty ways to produce profound marginalization" (1994: 897).

At present, sexual violence data is produced by a range of authorities, actors and technologies; to this end, I use bioinformation to map new terrain for feminist approaches to bioinformation studies. There's an emerging subset of feminist technoscience studies exploring the cultural-material practices of forensic science and rape response (Shelby 2018, Quinlan 2017, Kruse 2015). One such study is Shelby's (2018) examination of how "epistemological friction[s]" emerge in sexual assault treatment. Shelby (2018) explores "complementary and contested" practices that specifically arise in medico-legal contexts (Shelby 2018: 5). Using the example of the Vitullo® Kit, the prototype rape kit used to gather and preserve forensic evidence, Shelby (2018) argues that the kit is produced simultaneously by positivist criminology and protocol feminism; two knowledge positions seemingly at odds. By unpacking these epistemologies, Shelby (2018) shows how the kit is mobilised differently according to perceived value.

Quinlan provides a complementary analysis in *The Technoscientific Witness of Rape: Contentious Histories of Law, Feminism, and Forensic Science*, arguing that the rape kit is the "quintessential *modest witness* of rape" (2017: 17). Quinlan's history of the kit explores "witnessing" as a phenomenological encounter, during which the kit retains "a memory of the injuries and bodily fluids" left by the perpetrator(s) (2017: 15). Collected by medical staff, passed on to police, and viewed by lawyers and judges in the courtroom, the kit and its contents are consistently valued as credible objective facts (Quinlan 2017). Quinlan (2017) and Shelby (2018) both emphasise how rape kits (and other technoscientific interventions) have become stabilised as neutral truth-tellers. Like Quinlan and Shelby, I'm interested in how rape response is shaped by various epistemic cultures, feminist or otherwise. However, their work focuses more intensely on the rape kit and object-oriented technoscience. The kit may collect certain types of bioinformation, but this tool is not the only method to generate this data.

For this reason, I turn towards Kruse's definition of materiality in my discussion of bioinformation and sexual violence. As Kruse states in her own analysis of forensic evidence, sex and gender, *materiality* should be viewed as "an activity rather than a quality" (2010: 374). Knowledge materialises various meanings including "bodies and persons" (ibid.). These bodies are not biological, but instead are constituted by social relations and interactions. Kruse's arguments, like most technoscience scholarship, are influenced by Donna Haraway (2018), who introduces this concept of bodies:

> Technoscientific bodies, such as the biomedical organism, are the nodes that congeal from interactions where all the actors are not human, not self-identical, not "us." The world takes shape in specific ways and cannot take shape just any way; corporealization is deeply contingent, physical, semiotic, tropic, historical, located. Corporealization involves institutions, narratives, legal structures, power-differentiated human labor, technical practice, analytic apparatus, and much more. ... For humans, a word like gene specifies a multifaceted set of interactions among people and nonhumans in historically contingent, practical, knowledge-making work.
>
> (n.p.)

Haraway's example of the gene is useful for thinking through the material-discursive practices of bioinformation. In a similar sense, bioinformation is multifaceted, acting as a mediator of sorts between biological systems, data science, forensic knowledge and digital technologies. Rape response uses bioinformation to simultaneously assemble and disassemble bodies across carceral policies and practices. This chapter focuses on bodies through two relations: evidence and information. With public scrutiny turning to forensic commissioning and criminal justice outcomes, it is imperative to consider not only *how* bioinformation is implicated in sexual assault cases but *what/who* it bodies forth—and this must be of critical feminist interest.

Bodies-as-evidence: bioinformatic use and forensic "woundedness"

Ideas around gender norms, modesty and biological essentialism drive early representations of sexual violence. Definitions of sexual assault in the UK date back to Celtic law (around 1000 to 55 BCE), which recognised two types: forcible rape and rape of a woman who couldn't consent due to mental illness or intoxication (Bourke 2015, Sanyal 2019). Rape trials from the 16th century into the 19th century focused on two determinants: firstly, the theft of the victim's honour (if there was any "honor" to be stolen in the first place) and secondly, the intensity of the victim's physical resistance (Bourke 2015, Sanyal 2019). Long before the rise of forensic science, bodies and perceptions of "one's own vulnerability" dominated discourses around sexual violence, enforcing the perceived "difference between bodies in need of protection (vulnerable) and protecting/controlling bodies (invulnerable)" (Sanyal 2019: 121).

This perception of (in)vulnerability is racialised, reinforcing capitalist notions of private property and ownership of gendered bodies (Sanyal 2019; Phipps 2020). Freedman argues in *Redefining Rape: Sexual Violence in the Era of Suffrage and Segregation* that historic understandings of rape focused on attacks against "chaste, unmarried, white woman by a stranger" (2013). Even though Freedman looks at trials dating back to the 1700s, her analysis of historic rape law and the influence of slavery and colonialism is salient today. Take for example, Steenberg's (2013) description of forensic science and sexual violence in media: here again, we see perceptions of race, gender, class and ultimately "purity" being reproduced.

Some would argue that technoscientific tools like the rape kit seek to obfuscate this vicious culture of disbelief and subjugation; as I've outlined, forensic science is often celebrated for its apparent objectivity or truth telling (Shelby 2018, Quinlan 2017). However, time and time again perceptions of survivors' (in)vulnerability are resurrected in police investigations and courtrooms, whether to discredit or even criminalise victims. In fact, in the early stages of treatment and investigation, bioinformation is used to determine vulnerability as part of what Maguire and Rao call the "body-evidence relation" (2018: 4).

Maguire and Rao define this relation as the production of evidence and the subsequent recognition, classification and management of human life (Maguire and Rao 2018). Much like Kruse's (2010) articulation of forensic knowledge as an activity that materialises bodies, Maguire and Rao argue "what we know of a person is often the outcome of processes" (2018: 10). Maguire and Rao pay particular attention to migrant bodies and the use of bioinformation in border enforcement (i.e. biometrics), which they suggest is closely linked to the "contemporary drive for (and obsession with) forensic knowledge" (2018: 14). Similarly, for sexual assault investigations, the forensic script is a "reimagining of the social through the body," and bioinformation becomes a data capture for physical violation and psychological trauma (Maguire and Rao 2018: 10).

But what does bioinformation "look" like in the context of sexual assault? As Parry and Greenhough describe, bioinformation falls into two categories: derivative and descriptive. In the case of sexual assault, derivative bioinformation includes DNA evidence (i.e. bodily fluids, hair and skin) obtained using sexual assault kits in the acute aftermath of an assault. Derivative bioinformation is subsequently produced through the collection of organic substances and testing to identify a DNA profile. In comparison, descriptive bioinformation includes medical and forensic records—or archives—that translate the "fleshier" attributes of derivative bioinformation into searchable and readable databases (Parry and Greenhough 2017). I'd add that forensic medical examinations produce additional descriptive bioinformation, such as body maps for injury identification and digital recordings of internal exams.

When scholars discuss sexual violence and bodies-as-evidence (or the body-evidence relation) it's often in reference to forensic medical examinations and acute sexual assault services. Corrigan's extensive research on medical responses to sexual assault and abuse provides an exemplary analysis of evidence collection. Exploring roles such as SANEs, Corrigan details the jurisdictional tensions between police, medical specialists, hospitals and victims. In her words, post-assault examinations become a "trial by ordeal," in which survivors are figured as cooperative/uncooperative, sincere/insincere and credible/non-credible (2013: 132). Mulla refers to the expectations placed on survivors as a "corporeal discipline" (2014: 40). In *The Violence of Care: Rape Victims, Forensic Nurses, and Sexual Assault Intervention*, Mulla writes,

> Our common wisdom holds that if a rape victim submits to the scrutiny of a forensic examination, and organic substance is recovered, subsequent tests will reveal the DNA hidden in the evidentiary depths, and the identity of a perpetrator will be unambiguously indicated, putting justice more easily within reach.
>
> (2014: 37)

Corrigan and Mulla's analyses will not be accurate for all survivors who present at acute sexual violence services; research indicates specialist services like SARC or SANE programmes can greatly improve survivors' experiences. However, Corrigan and Mulla describe the context in which bioinformation makes the first part of its transition from biological substance (or other material evidence) to information. So much is negotiated during these moments, when clinicians interact with survivors, and evidence collection kits are introduced as essential tools. What is not often discussed during FMEs are the disparate outcomes of derivative bioinformation; consenting to a forensic medical examination involves consenting to an unknown.

For example, data from the Victims' Commissioner indicates that in 2019 there were 63,666 reported cases of rape; only 8% were referred to the Crown Prosecution Service (HM Crown Prosecution Service Inspectorate 2019). Another study shows that while biological evidence is the primary type of evidence collected, only 10.3% of sexual assault kits received a complete examination (Johnson, Peterson, Sommers and Baskin 2012). Findings suggest that the analysis of kits largely depends on police procedures: in only 1.6% of cases biological evidence was analysed prior to arrest (Johnson, Peterson, Sommers and Baskin 2012). Using DNA/STR techniques, for 10 out of 602 cases an association was made between victim and suspected perpetrator; in 0.7% of cases a match was made using a DNA index system (Johnson, Peterson, Sommers and Baskin 2012). In other words, bioinformation as a forensic device is not consistently processed and even when it is, the data is not always conclusive.

This transition from biological trace to actionable information is best described as a "chain of transformations" (Kruse 2010: 366). At each juncture, medical, forensic and legal experts track and assess bioinformation by its "usefulness" (Maguire and Rao 2018: 7). However, usefulness is largely absent from public discourses around bioinformation and forensic evidence. The "use" of bioinformation is much more subjective than public debates would admit; instead, bioinformation is presented by sexual assault services and criminal justice systems as a means to "establish clarity, certainty, and stability" through science (Code 1991: 50). This desire for forensic truth is not only a political issue but also an emotional one. As Manshel writes in *The Desire for Fact: Anti-Racist Ethics in Discourses of Sexual Violence*, "In discourses of sexual violence, victims and witnesses are expected to perform affect as evidence of *woundedness*" (2018: 511; emphasis added).

Similar to Sanyal's (2019) discussion of narratives of vulnerability, Manshel interrogates the "obligation for a rape victim to prove ... that she 'wasn't asking for it'" (2018: 516). Manshel further challenges liberal narratives of rape as a "personal problem" (2018: 517). When rape is over-individualised, Manshel (2018) argues that lines are drawn between right and wrong feelings. Forensic knowledge has rendered bioinformation as a kind of sterile, opaque object, yet body-evidence relations traffic in emotions. One survivor recalls in Quinlan's book *The Technoscientific Witness of Rape: Contentious Histories of Law, Feminism, and Forensic Science* the way in which descriptive bioinformation is collected,

> They ask if I take medication, have attempted suicide, been admitted to a psychiatric hospital. If I have had abortions, recent consensual sex ... then they ask me to tell my rape. They write everything down, record the data on forms with numbers and codes that have been waiting for me to be raped.
>
> (2017: xi)

As Jane Doe describes, she must tell her rape through a dialogue that classifies and evaluates her woundedness. For Manshel (2018) the performance of woundedness carries far beyond forensic encounters. From courtrooms to the memoir genre, Manshel draws upon numerous arenas wherein survivors are expected to perform and endure. For Manshel (2018), rape narratives and interventions that deal in "vulnerability and compulsory affect" perpetuate racist renderings of victimisation. Non-white and non-middle-class women are subject to more doubt, particularly when they are seen as "insufficiently traumatized" (2018: 516). Manshel's reading of sexual violence and trauma can be used to challenge static conceptualisations of bioinformation in two important ways. First, it animates bioinformation across time and place. Secondly, it demands a conversation about bodies and persons that recognises sexual violence as an extension of state power.

In terms of animating bioinformation, during an FME there are verbal, physical and material exchanges that produce derivative and descriptive bioinformation. Borrowing from Manshel, producing bioinformation demands not only documentation of the "injurious" in a literal sense but also documentation of "emotional victimization" (2018: 516). In the context of bodies-as-evidence, bioinformation is synonymous with a wound—it opens up a body to contamination or healing. There are also various degrees of bioinformation's potential injury, sometimes it simply is superficial. To view bioinformation in this light, however, requires us to explore bioinformation's transition/transformation into data. It's here that we encounter figurations of the raped body in national biorepositories, police databases, big data solutions and digital platforms.

Bodies-as-information: the data worlds of sexual violence

In 2008, the Nuffield Council on Bioethics released a report on the ethical issues of bioinformation. The report evaluates policing and criminal justice practices in the UK that relate to the collection, interpretation and retention of bioinformation (Nuffield Council on Bioethics 2008). Of principle concern for the report is the National DNA Database (UK NDNAD), which retains biological samples and DNA profiles. Emphasising agency, the council makes recommendations to change consent practices for victims, witnesses and arrestees. Addressing bioinformation from a liberal rights-based approach, the report highlights the need for greater attention to "liberty, autonomy, privacy, informed consent and equality" (Nuffield Council on Bioethics 2008: xiii).

Prior to the enactment of the 2012 Protection of Freedoms Act, consent given to retain samples and profiles in the UK NDNAD was irrevocable (Amankwaa and McCartney 2019). Retention is now limited to a maximum of six months for most profiles (Amankwaa and McCartney 2019). While reform introduced database restrictions, the council's report highlights practices in the retention of bioinformation; foremost that this is adjudicated by the "crime control potential" of the data (Nuffield Council on Bioethics 2008: xvi). Bioethics often falls short of nuanced analysis of bioinformation, concentrating instead on a normative balance between public good and personal interest. Much like backlog campaigns, which see forensic interventions as necessary for public safety, conversations around biological databases and archives espouse ideals for improving best practice across police and forensic authorities.

The UK NDNAD is one example of where bioinformation "lives" and how sexual violence data is governed, in this case by the Home Office. As property of the database, bioinformation accrues different values and meanings, particularly when it is levied as a resource for state security (see biometrics). The previous section dealt more conventionally with bioinformation as a product of forensic tools but bioinformation is much more multi-purpose and interconnected than a swab during an examination. This section shifts from bioinformation collection to data retention, storage and circulation; expanding on Maguire and Rao (2018), I call this the body-information relation. In order to better situate bioinformation within the politics of sexual assault cases, I look at what Hong refers to as "data-fication" or more specifically, the relationship between the data world(s) of sexual violence and technoscientific corporealisation (2020: 2).

Sexual assault, rape and sexual abuse are of interest to multiple authorities, including health, forensic, psychosocial, policing and legal expertise. However, the ways in which institutions define rape suggests a singularity—both in terms of incidence and embodiment. For this reason, criticisms of databases, such as the UK NDNAD, tend to participate in "single-issue rights-based campaign[s]" (Bumiller 2008: 164). It's suggested that individual consent will mitigate any potential harms of data collection, storage, analysis and use. The 2012 Protection of Freedoms Act is one such example of a rights-based argument, rewriting exactly what donors (including victims, witnesses, suspects, arrestees and so on) are consenting *to*; in this case, consenting to temporary ownership and eventual destruction of bioinformation.

In *Data Feminism*, D'Ignazio and Klein call the construction of subjects and settings a "data ecosystem" (2020: 170). The UK NDNAD exists within much larger data ecosystem of interactions and relations, wherein what can be stored, retrieved, copied, transferred, rewritten, circulated, accumulated and transmitted is constantly re-determined (Wilcox 2015). The tension between what data "knows" and public discourses of sexual violence is further complicated by digital data. In Chun and Friedland's discussion of new media and the non-consensual online sharing of nude or explicit images/videos, such as revenge porn or celebrity hacks, they recall figurations of sexual assault and victimhood.

> The traditional idea of female virtue—one that is destroyed by sexual experience or physical exposure—positions ideal female sexuality as contained, private, and invisible. The positioning of slanegirl and others as ruined suggests how the leak—not the sexual act per se—destroys the virtue of its victims. Both the notion of a leaky opening (slut) and of a violently penetrated interior (rape victim/ruined virgin) depend upon the promise of closure, of being sealed. This desire to contain female sexuality, to uphold the virtue of virginity, now plays out both in our orifices and our interfaces.
>
> (2015: 9)

Chun and Friedland echo the "logic of ruin" that has permeated cultures for centuries; we are now encountering novel domains of constructing sexual subjects (2015: 15). From top-to-toe examinations to online habits, these converging fields of data are utilised in sexual assault cases. Hlavka and Mulla (2018) outline the ways in which text messages are used during rape trials alongside forensic evidence derived from FMEs. Through a comparative case analysis, Hlavka and Mulla show how digital data is trapped within the "lacunae of 'he said, she said'" and other common rape tropes (2018: 433). Hlavka and Mulla's (2018) study reflects what Chun and Friedland call a "politics of memory as storage," through which our digital—and sexual—acts are "not just *out there* ... but out there *forever*" (2015: 15–16).

Contemporary investigatory processes have largely monopolised the disintegrating boundaries between physical and digital subjects. In the UK, investigations into cases of sexual assault, abuse and rape involved the seizure of physical devices (i.e. phones, tablets and computers) as well as unrestricted access to social media accounts and cloud storage. Stafford statements are by design all encompassing; they have also been used to collect descriptive bioinformation such as social care, educational, psychiatric and medical records. Digital data and personal records fall under the jurisdiction of both police and prosecution services. Official guidance suggests there should be reasonable scope for requesting any type of information, however an independent review revealed that the Crown Prosecution Service often rejects sexual assault cases that *lack* these comprehensive data profiles (Angiolini 2015).

In addition to data collection procedures, the Metropolitan police have begun trialling artificial intelligence software for the review of digital evidence; no exceptions are currently made for sexual assault victims' data (Big Brother Watch 2018, Bowcott and Barr 2020). Authorities argue that the use of AI software is about efficiency, implying again that it's in the public interest to modernise institutional approaches to data management and analysis (Bowcott and Barr 2020). Richterich states that there is a "common framing of big data approaches as 'unobtrusive'" (2018: 76). In the case of sexual violence investigations, AI is presented as less invasive, reiterating epistemological claims that forensic technologies are objective *witnesses*.

Critics of Stafford statements and the use of AI make two arguments against data-focused investigations: (1) consent is defined too broadly and may be used to collect irrelevant information and (2) victims are expected to comply with requests, deterring them from reporting or engaging with the criminal justice system (Bowcott and Barr 2020). A statement issued by Rape Crisis England and Wales (2018) describes how Stafford statements are commonly used:

> Through their frontline support work, Rape Crisis Centres have encountered examples of adult victims of recent rape having their school records scrutinised; how can this be deemed relevant or proportionate? In other cases, the time frames for evidence disclosure seem to be chosen at random; how can it be justifiable to trawl a complainant's entire digital

communications for a three-year period, for example, when the issue in question is whether they consented to a specific sexual act on one particular occasion?

Rape Crisis invokes a victim's right to privacy as well as the privacy of friends and family. In the same statement, Rape Crisis refers to data's "obvious sensitivity," addressing the intimate nature of personal records and digital data, and fears over being perceived as (in)vulnerable complainants (Rape Crisis England & Wales 2018). This description of data—defined by its sensitiveness—stands in contrast with how police and CPS frame their investigative procedures, preoccupied instead with tackling data's vastness. The statement concludes by citing multiple and persistent failures of the criminal justice system: lack of scrutiny of the accused, length of criminal investigations, and underfunding of specialist support and advocacy. Considering the history of anti-rape reform, it's possible that Rape Crisis is calling for a new "paradigm of reform," one that seeks to delimit the use of digital data and big data analytics but still reinforces carceral-feminist interests (Caringella 2009: 9).

However, the body-information relation is somewhat unfamiliar territory for the mainstream anti-rape movement. It's difficult to imagine what reform looks like amidst streamlined modes of "'seeing' into our bodies," which render victims as flows of information (Wilcox 2015: 115). These information flows also extend far beyond police databases, courtroom records and other spaces that seemingly administrate sexual violence data. A recent super-complaint submitted by Liberty and Southall Black Sisters identifies a deliberate lack of guidance around data sharing between all 43 police forces in England and Wales and the Home Office. The organisations argue that currently no legislation is in place in the UK to protect survivors with irregular immigration status; disproportionately this affects women of colour who are sexual and domestic violence survivors (Liberty and Southall Black Sisters 2018).

The super-complaint argues that data obtained by police authorities is used in "the process of separating out 'good' victims and 'bad' victims of crime" (Liberty and Southall Black Sisters 2018: 4). Police may share personal data with the Home Office to determine is a person is undocumented, which "takes place routinely in relation to BME victims and witnesses and/or those with accents" (Liberty and Southall Black Sisters 2018: 46). The ability to not only collect bioinformation and personal data from sexual violence survivors but to circulate the data is an essential immigration enforcement technology, distinguishing between "livable lives" and "bodies that don't matter" (Wilcox 2015: 201). Statements made by organisations like Liberty and Southall Black Sisters and Rape Crisis are equally concerned with this distinction and how data materialises rapeable and non-rapeable bodies.

The normalisation of sexual assault and abuse through data, and in particular the subjugation of non-white bodies, is not limited to the criminal justice system. Noble argues that our new data realities are shaped by the hegemony of search engines like Google, which refract "patriarchy, racism and rape culture" (2018: 94). This desire to build increasingly larger bodies of evidence—to have multiple forms of derivative and descriptive information accessible and archivable—demonstrates the punitive aim of mainstream rape response. While the connection between biological trace and algorithmic models is not immediately evident, my discussion of the body-information relation identifies how bioinformation infrastructure, such as data surveillance, reproduces these modes of power.

Surviving in the ether: towards a feminist analysis of bioinformation

A significant project of mainstream anti-rape reform is increasing access to forensic evidence collection. Services like SANE programmes and SARCs were developed as meeting grounds for advocates, clinicians and police. When scholars write about rape and technoscience, discussions generally revolve around survivors' experiences with these institutional actors, tools and practices. During the collection of derivative and descriptive bioinformation, survivors embody a multiplicity—patient, victim, crime scene, witness, self-advocate and so on. These experiences are complicated; Quinlan (2017) encapsulates it best by subtitling her book on the subject as *contentious* histories.

Objects like the sexual assault kit lend further insight into the "epistemological friction" between feminist advocacy and the rape script (Shelby 2018: 5). Even though the criminal justice system has never been a neutral stage for the politics of gender-based violence, there seems to be a consensus across medico-legal authorities that forensic knowledge will answer for both interpersonal harm and sexual violence as sociocultural phenomena. In many ways, the development of medico-legal interventions has strengthened liberal feminism's role in rape response. Yet several decades after the introduction of the prototype kit and programmes that facilitate its use, new issues have emerged in the UK around backlogs and forensic commissioning. This chapter revisits the embodied dimensions of the making of technoscientific bodies at the interface of medical, legal and criminal justice processes and systems of bioinformation collection, analysis and management.

Like Haraway's (2018) articulation of the gene as a set of interactions, bioinformation is similarly discursive. I've argued that in terms of the body-evidence relation, bioinformation is constituted as injury and wound. During FMEs, the examining doctor or nurse records physical injuries and collects biological trace as a kind of catalogue of violence that is transmissible as information. Through these interactions, bioinformation acts as a "transposition of rape as sexual intercourse," recreating sexual violation for classification and analysis (Hartman 1997: 85). As an affective encounter, survivors also perform their woundedness; again, bioinformation becomes a record of traumatisation that may be used by police, prosecutors and other observers of data to draw conclusions about (in) vulnerability.

Beyond bioinformation collection, the body-information relation problematises the datafication of criminal justice systems and state security—or the "sticky web of carcerality" (Benjamin 2019: 2). While DNA databases are the most obvious examples of bioinformation management, the turn to data-driven policing means that our digital selves have become another form of evidence. Anti-rape organisations attempt to show how DNA retention, Stafford statements, algorithms and data sharing are unethical, and to humanise survivors. In *Carceral Capitalism*, Wang explores how racial capitalism shapes perceptions of personhood and innocence. She writes,

> When innocence is used to select the proper subjects of empathetic identi-fication, it also regulates the ability of people to respond to other forms of violence such as rape and sexual assault. ... "Promiscuous" women, sex workers, women of color, women experiencing homelessness, and people addicted to drugs are not seen as legitimate victims of rape ... Rape is often conventionally defined as "sexual intercourse" without "consent," and consent requires the participation of subjects in possession of full person-hood. *Those considered not-human cannot give consent.*
>
> (2018: 288)

Wang argues against insisting on innocence as political praxis: "When we challenge sexual violence with appeals to innocence, we set a trap for ourselves by reinforcing the assumption that white cis women's bodies are the only ones that cannot be violated" (2018: 291). Wang's arguments "against innocence" mirrors Sanyal's (2019) discussion of vulnerability. They both show that vic-timhood is figured by discourses of chastity (the existence of purity to be stolen by rape) and virtue (an indelibility of character that is legible by medical and legal expertise). The *trap* that Wang warns of is an adherence to the "limits of law" and by extension, the forensic technologies that seek to describe sexual assault and abuse (2018: 291).

In this chapter, I've also teased the following question: by participating in state-led rape response, did mainstream feminism open up survivors to data trawls like Stafford statements? By no means can I answer this but it's my hope to show that public anxieties over forensic provisions play into the surveillance state, pulling racialised bodies into dangerous focus. With more data circulating for use by law and immigration enforcement, Davis and Dent's words feel immediate: "the prison is itself a border" (2001: 1236–1237). As a study in/of bioinformation, this chapter invites both bioinformation studies and feminist technoscience to view sexual violence data (whether biomedical, forensic or digital) as material-cultural practices. These practices silo meaning from skin to database.

Technoscience and its relationship to rape response seems like an endless discourse of suffering. But there *is* a possible future where anti-rape feminism comes to terms with its denial and divide. Bioinformation—framed in this volume as relations and interactions—shows us that these futures must reveal themselves in haste. To welcome this, sexual violence scholarship should engage in knowledge-making work that centres marginal subject positions. Across sociology and anthropology, sexual assault needs to be grounded in the social and systemic conditions that create cultures of violence (Baker and Bevacqua 2018). Scholars and activists can learn from bioinformation and its ability to convey collective knowledge: "a singular action [is] never singular, because it is linked to a pattern elsewhere" (Chun 2016: 374). Trauma like data speaks of things obscured and of metrics for resistance.

References

Amankwaa, A.O. and McCartney, C. (2019) 'The effectiveness of the UK national DNA database', *Forensic Science International: Synergy*, 1, pp. 45–55.

Angiolini, E. (2015) Report of the Independent Review into The Investigation and Prosecution of Rape in London. Available at: www.cps.gov.uk/sites/default/files/documents/p ublications/dame_elish_angiolini_rape_review_2015.pdf (Accessed 7 March 2020).

Baker, C.N. and Bevacqua, M. (2018) 'Challenging narratives of the anti-rape movement's decline', *Violence Against Women*, 24 (3), pp. 350–376. doi:10.1177/1077801216689164.

Benjamin, R. (2019) *Captivating Technology: Race, Carceral Technoscience, and Liberatory Imagination in Everyday Life*. Durham, NC: Duke University Press.

Big Brother Watch (2018) Justice Committee: Disclosure of evidence in criminal cases inquiry. Available at: https://bigbrotherwatch.org.uk/wp-content/uploads/2018/07/Big-Brother-Wa tch-evidence-Disclosure-of-evidence-in-criminal-cases.pdf (Accessed 7 March 2020).

Bowcott, O. and Barr, C. (2020) 'Impact on rape victims of police phone seizures to be reviewed', *The Guardian*, 16 February. Available at: www.theguardian.com/society/2020/ feb/16/impact-on-victims-of-police-phone-seizures-to-be-reviewed (Accessed 1 March 2020).

Bourke, J. (2015) *Rape: A History From 1860 To the Present*. London: Hachette.

Boyer, A. (2011) *My Common Heart*. Louisiana and Texas: Spooky Girlfriend Press.

Bumiller, K. (2008) *In an Abusive State: How Neoliberalism Appropriated the Feminist Movement against Sexual Violence*. Durham, NC: Duke University Press.

Caringella, S. (2009) *Addressing Rape Reform in Law and Practice*. New York: Columbia University Press.

Centers for Disease Control and Prevention (2014) Sexual Violence Surveillance: Uniform Definitions and Recommended Data Elements. Available at: www.cdc.gov/violencep revention/pdf/sv_surveillance_definitionsl-2009-a.pdf (Accessed 15 February 2020).

Chun, W.H.K. and Friedland, S. (2015) 'Habits of leaking: of sluts and network cards', *Differences: A Journal of Feminist Cultural Studies*, 26 (2), pp. 1–28. doi:10.1215/ 10407391-3145937.

Chun, W.H.K. (2016) 'Big data as drama', *ELH*, 83 (2), pp. 363–392. doi:10.1353/ elh.2016.0011.

Code, L. (1991) *What Can She Know? Feminist Theory and the Construction of Knowledge*. Ithaca, NY: Cornell University Press.

Corrigan, R. (2013) *Up Against a Wall: Rape Reform and the Failure of Success*. New York: NYU Press.

Crenshaw, K. (1991) 'Mapping the margins: intersectionality, identity Politics, and violence against women of color', *Stanford Law Review*, 43, 1241–1299. doi:10.2307/1229039.

Davis, A. and Dent, G. (2001) 'Prison as a border: a conversation on gender, globalization, and punishment', *Signs*, 26 (4), pp. 1235–1241.

Devlin, H. (2019) 'Forensic science labs are on the brink of collapse, warns report', *The Guardian*, 1 May. Available at: www.theguardian.com/science/2019/may/01/foren sic-science-labs-are-on-the-brink-of-collapse-warns-report (Accessed 10 November 2019).

D'Ignazio, C. and Klein, L.F. (2020) *Data Feminism*. Cambridge: MIT Press.

Egan, S. (2019) 'Excavating feminist knowledges and practices in the field of sexual assault service provision: an Australian case study', *Women's Studies International Forum*, 74, pp. 169–178.

Freedman, E. (2013) *Redefining Rape: Sexual Violence in the Era of Suffrage and Segregation*. Cambridge, MA: Harvard University Press.

Gruber, A. (2007) 'The feminist war on crime', *Iowa Law Review*, 92, pp. 741–833.

Gruber, A. (2020) *The Feminist War on Crime: The Unexpected Role of Women's Liberation in Mass Incarceration*. Oakland: University of California Press.

Haraway, D. (2018) *Modest_Witness@Second_Millennium.FemaleMan_Meets_Onco-Mouse: Feminism and Technoscience*. London: Routledge.

Hartman, S. (1997) *Scenes of Subjection: Terror, Slavery, and Self-making in Nineteenth-century America*. Oxford: Oxford University Press.

Hlavka, H.R. and Mulla, S. (2018) '"That's how she talks": animating text message evidence in the sexual assault trial', *Law & Society Review*, 52 (2), pp. 401–435.

HM Crown Prosecution Service Inspectorate (2019) 2019 rape inspectorate. Available at: www.justiceinspectorates.gov.uk/hmcpsi/wp-content/uploads/sites/3/2019/12/Rape-insp ection-2019-1.pdf (Accessed 20 February 2020).

Hong, S. (2020) *Technologies of Speculation: The Limits of Knowledge in a Data-Driven Society*. New York: NYU Press.

Johnson, D., Peterson, J., Sommers, I. and Baskin, D. (2012) 'Use of forensic science in investigating crimes of sexual violence: contrasting its theoretical potential with empirical realities', *Violence Against Women*, 18 (2), pp. 193–222. doi:10.1177/1077801212440157.

Kruse, C. (2010) 'Forensic evidence: materializing bodies, materializing crimes', *European Journal of Women's Studies*, 17 (4), pp. 363–377. doi:10.1177/1350506810377699.

Kruse, C. (2015) *The Social Life of Forensic Evidence*. Berkeley, CA: University of California Press.

Liberty and Southall Black Sisters (2018) 'Police data sharing for immigration purposes: a super-complaint'. Available at: https://assets.publishing.service.gov.uk/government/uploads/system/uploads/attachment_data/file/767396/Super-complaint_181218.pdf (Accessed 20 January 2020).

Manshel, H. (2018) 'The desire for fact: anti-racist ethics in discourses of sexual violence', *Criticism*, 60 (4), pp. 511–531. doi:10.13110/criticism.60.4.0511.

Maguire, M. and Rao, U. (2018) 'Introduction: bodies as evidence', in Maguire, M., Rao, U. and Zurawski, N. (ed.), *Bodies as Evidence Security, Knowledge, and Power*. Durham, NC: Duke University Press.

Martin, P.Y. (2005) *Rape Work: Victims, Gender and Emotions in Organization and Community Context*. Hove: Psychology Press.

Mulla, S. (2014) *The Violence of Care: Rape Victims, Forensic Nurses, and Sexual Assault Intervention*. New York: NYU Press.

Murphy, M. (2012) *Seizing the Means of Reproduction: Entanglements of Feminism, Health, and Technoscience*. Durham, NC: Duke University Press.

Noble, S.U. (2018) *Algorithms of Oppression: How Search Engines Reinforce Racism*. New York: NYU Press.

Nuffield Council on Bioethics (2008) The forensic use of bioinformation: ethical issues. Available at: www.nuffieldbioethics.org/publications/forensic-use-of-bioinformation (Accessed 29 February 2020).

Parry, B. and Greenhough, B. (2017) *Bioinformation*. Hoboken, NJ: Wiley.

Phipps, A. (2020) *Me, Not You: The Trouble with Mainstream Feminism*. Manchester: Manchester University Press.

Quinlan, A., Fogel, C. and Quinlan, E. (2010) 'Unmasking scientific controversies: forensic DNA analysis in Canadian legal cases of sexual assault', *Canadian Woman Studies*, 28 (1), pp. 98–107.

Quinlan, A. (2017) *The Technoscientific Witness of Rape: Contentious Histories of Law, Feminism, and Forensic Science*. Toronto: University of Toronto Press.

Quinlan, A. (2019) 'Visions of public safety, justice, and healing: the making of the rape kit backlog in the United States', *Social & Legal Studies*, pp. 1–21. doi:10.1177/0964663919829848.

Rape Crisis England & Wales (2018) The real issue with evidence disclosure in rape cases. Available at: https://rapecrisis.org.uk/news/latest-news/the-real-issue-with-evidence-disclosure-in-rape-cases (Accessed 15 March 2020).

Razack, S. (1994) 'What is to be gained by looking white people in the eye? Culture, race, and gender in cases of sexual violence', *Signs*, 19 (4), pp. 894–923.

Richterich, A. (2018) *The Big Data Agenda: Data Ethics and Critical Data Studies*. London: University of Westminster Press.

Rojas Durazo, A.C. (2017) '"We were never meant to survive": fighting violence against women and the Fourth World War', in INCITE! (ed.) *The Revolution Will Not Be Funded: Beyond the Non-Profit Industrial Complex*. Durham, NC: Duke University Press, pp. 113–128.

Sanyal, M. (2019) *Rape: From Lucretia to #MeToo*. Brooklyn: Verso Books.

Shelby, R. (2018) 'Whose rape kit? Stabilizing the Vitullo® Kit through positivist criminology and protocol feminism', *Theoretical Criminology*, pp. 1–20. doi:10.1177/1362480618819805.

Steenberg, L. (2013) *Forensic Science in Contemporary American Popular Culture: Gender, Crime, and Science*. London: Routledge.

Topping, A. (2020) 'Police and CPS scrap digital data extraction forms for rape cases', *The Guardian*, 16 July. Available at: www.theguardian.com/society/2020/jul/16/police-and-cps-scrap-digital-data-extraction-forms-for-cases (Accessed 10 August 2020).

Wang, J. (2018) *Carceral Capitalism*. Cambridge: MIT Press.

Wilcox, L.B. (2015) *Bodies of Violence: Theorizing Embodied Subjects in International Relations*. Oxford: Oxford University Press.

7 American bioinformation and U.S. race politics

The values of diverse genetic data

Anna Jabloner

Introduction

Over the approximate decade I have spent listening to experts on biological information in the United States, I have registered a tenor in conversations and in biomedicine's public articulations:[1] *Bioinformation should reflect the diversity of all Americans*. In particular, medical genetic data do not currently, but should ideally, represent the diverse bodies in the U.S. in the form of biological information. Diversity characterizes the American population. Thus, it seems as simple as it does commonsensical, this diversity should be represented in biological information, signaling that biomedical research and its benefits are for everyone, and signaling also that new medical projects have the political potential to right historical wrongs. The National Institutes of Health (NIH)—primary funder of American biomedical research—for example writes about a U.S.-wide precision medicine initiative:[2]

> Diversity is one of the core values of the *All of Us* Research Program. *All of Us* is asking lots of people to join the program. Participants are from different races, ethnicities, age groups, and regions of the country. They are also diverse in gender identity, sexual orientation, socioeconomic status, education, disability, and health status.[3]

Other kinds of American bioinformation, however, seem incommensurable with this diversity-nation-morality matrix. In particular, the bioinformation collected in forensic domains into a federal genetic database, and used to identify persons or criminal suspects, is also expected to reflect the American population biologically. This bioinformation, however, overrepresents some of the U.S.'s racial and ethnic minorities, causing the federal database to be accused of racial bias. Because of this imbalance, experts frequently call for a universal DNA database, meaning that all people in the U.S., and ideally everyone on the planet, should be required to contribute a genetic sample. Forensic bioinformation could potentially contain all the diversity American biomedicine dreams of, yet the discussion in this context is not framed around diversity, but the value in genetic data lying instead in its being *universal*. Both

DOI: 10.4324/9780367810030-7

perspectives project promises to a national collective—marked or unmarked. Both originate in U.S. institutions of science and law whose enduring straight-white-male leadership is neither diverse nor universal, yet has long played the universalist "god trick of seeing everything from nowhere" (Haraway 1988: 581).

The notion of biological material as information holds inherent tensions.[4] Anthropologist Mike Fortun (2008), writing about genomics as large-scale generator of bioinformation, has productively theorized these tensions as chiasmus, or sets of seemingly contradictory terms that actually depend on one another. Bioinformation is necessarily situated because it comes from specific bodies, yet it is universal as a new global form that is usable across different domains. It is readable by computers as data, yet often opaque in its meaning. Importantly, bioinformation is individual in its provenance from a single body, yet assembled to represent some kind of collective—and this often happens in terms of race (M'charek 2000). In this chapter, I build on Fortun's notion of an inherent friction within genomics to argue that the *value* of bioinformation in the context of American race politics also emerges within these very tensions of specificity/universality and individuality/collectivity. Bioinformation and genetic data are not stable entities. Rather, diversity is varyingly ascribed to data across different racial, biopolitical, as well as biocapitalist logics in medicine and forensics.[5] The representation of specific groups also comes to matter differently across these domains.

In liberal America, diversity is generally seen as valuable.[6] Referring to value(s), I draw on anthropologist Hadas Weiss' productive theorization of values' nature as a specifically capitalist form of normativity, one that "presuppose[s] a likeminded community" (2015: 243). Weiss claims that values are interesting not primarily for their indication of morality, but for "the work they perform socially" (2015: 251). Thus, I ask what it means that diversity is a core value of medical research. Beyond the everyday meaning of value/s as something good or important, I draw on Weiss' conceptualization to illuminate how the value of diverse bioinformation enmeshes moral with economic rationales in the projects I discuss in this chapter. Diversity as a value propels or animates many of the data practices at stake, but value itself is already enmeshed within the current socio-economic context of the contemporary United States.

I approach an anthropology of bioinformation in terms of the ethical and political questions embedded in such artifacts, the values that this information accrues as it travels from bodies to databases and, sometimes, across legal, biomedical, and consumer contexts. The chapter proceeds by contrasting instances in which diversity in genetic data is valued with those in which it is not. Asking what and whose values are reflected in demands that bioinformation from consenting research subjects reflect diversity, I first articulate critical conceptualizations of diversity in American science and politics. I theorize anxieties regarding diversity as a way to interrogate articulations of fairness and challenges to the enduring American racial order in frameworks of biocapital.

I then engage an ongoing discussion, in the idiom of the universal, about expanding the federal U.S. DNA forensic database (CODIS) into a nation-wide database. I lay out this example to show that CODIS engenders an older but ongoing moment of U.S. race politics: colorblindness (Bonilla-Silva 2006).[7] This notion is reflected in the fact that the bioinformation stored in this database, from non-consenting subjects of arrest or conviction, includes only non-coding or so-called "junk" DNA. These data serve the legal identification of individuals grounded in the probabilistic matching of allele lengths, and are not understood as embodiments of racial or other identities. In what follows, I contrast CODIS with a number of examples that illustrate how diversity claims in STEM bolster an American-made genetic universalism. This genetic universalism blends liberal demands for integrating racial difference with epistemologies of molecularly embodied identities and the prospect of diversity as economic advantage (Fullwiley 2014, Shankar 2015). I elaborate on the *All of Us* program to draw out a connection between the value of diversity and the idea of personalization in medicine. Finally, I examine the notion of consent across the cited projects. This key bioethical term adds to the political and economic value of diverse bioinformation in medicine, but is absent in forensics. Since forensic bioinformation, with some exceptions, does not constitute consented data, I speculate that such data cannot be diverse in the first place. Grounded in different epistemologies of DNA and U.S. race politics, I conclude that competing universalisms characterize genetics in a moment of its totalizing expansion across social domains.

Why diversity? Critical takes on American biopolitics

From the 1960s onward, mass social movements, in particular ethnic and women's movements, organized to change a racist and sexist American status quo. Student activism played a major role in the societal upheavals of this time, resulting in the founding of ethnic and gender studies programs at colleges and universities, and thus initially bringing diversity into the academy through activism. New paradoxes of institutionalization ensued as these and other new inter-disciplines challenged, but were also domesticated into, traditional university structures (Ferguson 2012, Lowe 1996, Subramaniam 2009).

The field of feminist science studies originates with these social protests as women's movement activists studying natural sciences in the U.S. pushed back against the one-dimensional scientist, homogenous collectives, and oppressive structures they encountered in their respective disciplines (Harding 1986, Keller 1982, Subramaniam 2000). In consequence, some activists entered university administrations and governmental structures such as the NIH and the National Science Foundation (NSF) to overcome racist, sexist, and other oppressive institutional cultures, and ultimately to change scientific practices (in various functions preceding today's common "diversity officers"). And others built careers within science itself, in and across the traditional natural science disciplines, emerging women's studies programs, and within feminist science studies. Across these settings, activists and scholars challenged andro- and Euro-centric presuppositions across all academic knowledge enterprises, a project arguably also at the heart of current calls to decolonize the academy (Ahmed 2012, Fausto-Sterling 1992, Hammonds and Subramaniam 2003, Subramaniam 2014).

Longstanding political struggles to intervene on histories of inequality in science precede colorblindness as much as diversity management and the current commonsense notion of racial and gender inclusion across academy and industry. Feminist theorist Sara Ahmed (2007, 2012) argues that the concept of diversity served to give new life across Anglo-American universities to tired demands for equality in institutional settings. And yet, the new term also shed this history of political struggle, a history that, as Ahmed shows, those working toward diversity in different academic institutions now constantly have to bring back into view through their practices.

Diversity is a value that should ideally be articulated not just in biomedical researchers but within the data these researchers collect from people, within bioinformation itself (Reardon 2017). The NIH reacted to such political struggles by building new institutional mechanisms intended to effect changes in the homogenous white-male scientist demographics, for example through funding streams to include and retain women and racial-ethnic minorities in professional trajectories. In 1993, the NIH Revitalization Act also made binding "the inclusion of women and minority groups in all NIH-funded clinical research in a manner that is appropriate to the scientific question under study" (NIH n.d. a). Crucial in this domain were the struggles of patient activists who demanded to be included in research as populations—as bodily sources of important biological information—to ensure that outcomes, such as new treatments, would benefit them as affected groups (Harrington 2008). Women's and ethnic movement activists likewise often demanded inclusion as research populations, arguing that the standard biomedical white-male body and engendered one-size-fits-all kind of medicine needed to be overcome to alleviate ingrained U.S. health disparities (Epstein 2007, 2008). And yet, medical sociologist Steven Epstein (2007) argues that under the guise of inclusion of diverse minorities in biomedical research, attention also shifted from social inequality to biology. This is particularly detrimental in the current moment in U.S. race politics when, anthropologist Duana Fullwiley writes, the "acceptance of race as genetic is becoming ever more entrenched in medicine, law, science, education, genomic research, and personal identity" (2014: 814), just as diversity has become a stand-in for the default invisible whiteness of largely unchanged institutions (Rosa and Bonilla 2017).

The NIH embraces the inclusion of diverse researchers and research populations at a time when, on a larger political scale, diversity has also been reframed as a key aspect of economic growth and neoliberal policy (Urciuoli 2011, 2016). Analyzing former U.S. president Obama's State-of-the-Union speech in January 2013, political theorist Wendy Brown argued that his praise for a range of political demands such as equality and fairness framed these demands in terms of their "contribution to economic growth or American competitiveness" (2015: 25). In economic frameworks of innovation in American industry, diversity is considered to foster creativity (Hewlett et al. 2013). For the NIH, diversity has likewise become a key unifying value that signals combined potentials for improved health, creativity, and a growing U.S. economy (Jabloner and Lee 2020). Such trajectories are embedded in longer histories of subsuming STEM research under national economic development policy (Slaughter and Rhoades 2004). Rather than demand equality or complain about injustice, current innovation and industry speak praises the economic dividends resulting from diversity—but perhaps strategically so, in a language that can be adopted by anyone. To be sure, scholars of diversity urge us not to simply dismiss efforts to diversify. Ahmed (2012) for instance calls for empirical investigations of the work diversity does. Arcienega (2019) criticizes diversity and inclusion's anchoring in a neoliberal economy, but insists on these terms' potential as real instruments of institutional change.

In the U.S., diversity is a buzzword with particular political valences and long histories. Today, as ongoing lawsuits continue to play out a cruel backlash against affirmative action meant to mitigate the societal inequality that was built into the formation of the U.S. state, Ahmed reveals this notion's paradoxes and reductionist potentials. She writes that diversity as "an equality regime *can be* an inequality regime given new form, a set of processes that maintain what is supposedly being redressed," and that it "can participate in the creation of an idea of an institution that allows racism and inequalities to be overlooked" (2012: 8, 14. Emphasis in the original). Ahmed considers how diversity can reveal or conceal "institutional whiteness," which is usually the background against which diversity work becomes necessary (2012: 33). Importantly, Lee et al. argue in the context of precision medicine that in current calls for diversity, key political "questions arise about how to describe, define, measure, compare, and explain inferred similarities and differences among individuals and groups" (2019: 941). Against the background of institutional whiteness, herein lie the dilemmas of how specific individuals come to represent groups and how representations of specific groups come to matter in different contexts contained in the term diversity.

It is from such perspectives that I approach the diverging values of diverse genetic data. In bioinformation, diversity can be a way to channel complaints about racial injustice, for instance in the lack of relevant medical treatments for marginalized groups, into new practices such as demands for data to represent all Americans—as if transforming life into biodata could seamlessly translate into improved health outcomes without worlds in between. To have, or even just to demand a diverse project representative of Americans, on the other hand, can appear as evidence of institutional fairness but veil exclusionary histories. And yet, diversity is not the only value marker of bioinformation. In the forensic domain, bioinformation serves the purpose of legally identifying persons and is not used to embody identity or represent any specific groups other than a universal human collective. In what follows, I argue that parallel regimes of value apply to bioinformation across these domains, characterized by two key differences. The first difference consists in ideas about what diverse or race-neutral data are supposed to represent, the second in the consensual or coerced processes by which data are derived.

Colorblind bioinformation and calls for a universal DNA database

In the U.S. forensic domain, bioinformation predominantly comes from samples derived either from crime scenes or from buccal swabs taken from criminal offenders, and from persons who are arrested for specific felonies. This bioinformation is archived in the FBI-run CODIS database. While fairness is to be achieved in biomedical domains by representing diversity, and paying specific attention to bioinformation from groups that have been neglected in research, forensics engenders the notion that bioinformation is fair insofar as it represents, in an undifferentiated way, the whole of the nation or of humanity, and is thus race-neutral.

Historian Jay Aronson (2007) documented how decades of deliberation led to the identification of what exactly of the collected bioinformation—what kinds of genetic information—should be used in forensic applications. Short Tandem Repeats (STRs) were the agreed upon outcome, found in the non-coding DNA that makes up 97% of the human genome. This information has often been referred to as "junk DNA" because it does not code for any particular traits or phenotypes. At 20 genomic loci, selected to be biologically irrelevant, an individual's unique allele lengths are sequenced and compared across the database for matches. People are extremely unlikely to share these allele lengths at the defined locations, unless they are related to one another (Cho and Sankar 2004). CODIS data thus do not represent any established group but are grounded in a simple comparative logic.

CODIS is a criminological genetic database: it is not surprising that this archive overrepresents those U.S. racial and ethnic minorities who are arrested and convicted more often than others in a society rampant with racism. It is no surprise who is being remembered here, to draw on Geoffrey Bowker's (2005) notion of archives as indicative of what and how a society chooses to remember. Neither is it surprising that the U.S. does not advertise CODIS bioinformation as diverse. But the Black and Brown men whose bioinformation is disproportionately captured in CODIS are indeed the very demographics that are also disproportionately targeted in calls to participate in biomedical research, such as in the *All of Us* project cited above: they are the missing diversity of medical bioinformation.

Crucially, CODIS data aim for race-neutrality: genetic profiles are indexed here only by a number. Other information about a person, such as their race, ethnicity or gender, is meta-data in CODIS, meaning it is indexed separately from the bioinformation and not thought to lie *in* the bioinformation (Jabloner 2019, Murphy and Tong 2019). CODIS, however, reflects the biases of the U.S. criminal system: arrest and incarceration practices disproportionately target Black and Latino men, and their bioinformation is consequently overrepresented in CODIS. In turn, these groups' risks of becoming implicated in criminal investigations rise (Murphy 2015).

To repeat, STR bioinformation was selected because it does not signal traits or phenotypes. This idea of colorblind DNA crucially differs from the idea that diverse people's data should be used in biomedicine because their data themselves will reflect diversity, will reflect, in essence, race and ethnicity but also other signifiers as molecular identities (Fullwiley 2007). While massive debates continue in genetics and beyond about race and ethnicity as biological entities, the choice of STRs in U.S. forensics was grounded in the logic that while we all have these different allele lengths, they do not code for anything (Murphy 2018). Indeed STRs continue to operate as race-neutral identification-by-matching-numbers bioinformation in CODIS. These kinds of bioinformation are not just biologically irrelevant but understood to *apolitically* match individuals with crime scene DNA through probabilities when compared across massive databases (cf. the U.S. Supreme Court's characterization of junk DNA in Maryland *v.* King 2013). This colorblind epistemology, namely that data do not represent specific racial or ethnic groups, grounds the value regime of CODIS data as race-neutral data.

However, forensic scientists also use reference databases that are categorized by race and ethnicity, revealing the epistemic tensions that underlie this idea of neutrality (Cho and Sankar 2004, M'charek 2000). And what is more, genetic research on STRs itself presents profound epistemic puzzles about the very nature of genes and their consequential relationship with something like identity. Notably, anthropologist Nadia Abu El-Haj has analyzed geneticists' understanding of junk DNA as biologically irrelevant molecular signals that are however seen to contain historically embodied cultural practice as "silent signifiers of evolution" (2012: 48). From this perspective, Abu El-Haj argues, ideas about biological determinism are fundamentally changed, because it is cultural practice—known through historical sources—that determines the consequential biological signals we can read in contemporary bodies. On these epistemic grounds, genetic ancestry testing companies in the U.S. currently exploit the potentials of junk DNA for cultural significance, in what Abu El-Haj calls a "post-facto determinism" through which people can learn from genetic ancestry tests who they "have always already been" (2012: 247).

Regardless of the pending scientific truths about STRs, the idea in CODIS to create a neutral, not a diverse, database, and the subsequent realization that the database is *biased*, prompted a range of experts to call for a universal forensic genetic database at the national scale. Proponents of this idea argue that to ensure fairness, everyone should have a genetic sample taken upon birth, which is stored in a database for life (Hazel et al. 2018, Kaye and Smith 2004, Williamson and Dunkel 2002). For instance, bioethicists and legal scholars James Hazel et al. propose a universal U.S. genetic database, arguing that because this database would no longer be associated with crime, any social stigma of being in it would fall away. According to these authors, clearly defining what aspects of genetic information can be sequenced and stored would also improve the current situation in the U.S. where law enforcement can access all kinds of bioinformation across consumer and biomedical databases with relative ease.

Finally, they argue, everyone would be interested in privacy protections because their own information would be in the database (Hazel et al. 2018). Rather than challenge or support these claims, I state them to draw attention to the movement from race-neutral DNA towards calls for universality in a context where bioinformation disproportionately comes from socially marked groups. The call for universal bioinformation contrasts starkly with the logic of diverse, race-specific DNA in the biomedical realm, which I will outline in the next section.

The magical thinking of diversity in biomedicine

At countless conferences, I have heard geneticists lament what they call a "data disparities" problem in genomics. Journal articles evidence the idea that genomics' problem is its *data*'s lack of diversity (cf. Need and Goldstein 2009, Popejoy and Fullerton 2016, Sirugo et al. 2019), while news media state that "White-people-only DNA tests show how unequal science has become" (Regalado 2018). Mental health experts Stevenson and Atwoli (2019) write online that "[o]ne of the main issues presented by this diversity problem is that any solutions (including new medications) are likely to work best for the people whose DNA the research was based on – people of European descent." In this exemplary articulation of American race politics, a number of logics and values are in play. First, finding a solution to a health problem entails a range of social processes and crossover between domains that here remain veiled. Equally unclear, here as elsewhere, is the relevance of descent for the medical issue at hand: it is obvious for genetic diseases that cluster in geographic populations. Yet for an infectious disease, or a mental health disorder, it is unclear why the descent of the bioinformation would matter more than the actual circumstances of the individual it came from. Contrasting sharply with CODIS' race-neutral DNA, Fullwiley identified this problematic as "scientists and nonscientists deploy[ing] racialized genetics to address racial problems in society" (2014: 804). We should also ask how and under what circumstances health disparities would be assumed to flatten when genetic databases are more diverse, even if genes were actually the cause of the disparities (cf. Lee 2008, Reardon 2017).

Diverse biological data are here valuable bioinformation at multiple levels: for more socially representative science, research, and businesses, for potentially improved health outcomes from genomic research for wider social groups than white Americans—a group therein also projected as genetic population—or for the implied opening up of new consumer markets. Another angle of this data-disparities argument are renewed calls for genomics to "capture unseen diversity" across the Global South (Ulrich 2018). In urgent appeals for geneticists to study and sample genetically under-represented populations under the same trickle-down technoscience logic and presumption of molecularly embodied identity, human groups, at times entire continents, are framed as untapped potential and geneticists' goldmine in languages reminiscent of explorers and early colonial biologists (cf. Haraway 1989, Pratt 1992, Reardon and TallBear 2012).

Such calls to genetically capture diversity evoke the 1990s Human Genome Diversity Project (HGDP) whose plan to sample human biological diversity principally targeted indigenous populations around the globe (Reardon 2005). Otherwise left to their own devices, marginalized bodies took center stage as highly valuable bioinformation in a science project, yet then in a colonialist save-from-extinction, rather than the current biocapitalist trajectory. The HGDP was charged with bio-colonialism by indigenous collectives around the globe (Harry and Marks 1999). With the known histories of scientific (and in particular genetic) capture and extraction, as well as medical exploitation, the giving or taking of bioinformation signals that elements of bodies become value in someone else's enterprise and holds an intense place in the American collective memory. Ethnic, racial and sexual minorities, as well as poor people in the U.S., might rationally mistrust a biomedical apparatus that has profoundly impacted or violated their lives (cf. Garrison 2013, Lombardo 2008, Ordover 2003, Reverby 2009, Stern 2016, Washington 2006). This lack of trust is often cited as the reason for racial and ethnic minorities' under-representation in genomics, yet is presented as a problem to be overcome, for instance through the community liaisons of projects already underway.

In another instance, during a genetics and health disparities conference I attended in San Francisco, a famous Stanford geneticist exclaimed about recent Genome Wide Association Studies (GWAS):[8] "GWAS is run by Caucasians! ... We need to get minority data!" Diversity as a value here travels from attempts to make researcher demographics more heterogeneous to attempts to diversify the bioinformation they use. Again, this researcher connected the demographic non-diversity of *geneticists* conducting the studies with the non-diversity of the *bioinformation* in the databases. Diversity points not just to the marginalized populations who are missing in the databases, i.e. the subjects of bioinformation, but to its own invisible reference point, meaning the non-diverse, white American men who predominantly build those databases, which is to say, the producers of bioinformation.

The connection between the demographic change in scientists and in data is that diverse researchers are assumed to bring with them the potential for diverse data through social connections (cf. Popejoy and Fullerton 2016). Other presumptions are that again only with "their" data can biomedicine be made applicable to "them," and that trickle-down benefits of scientific advances for marginalized populations will ensue.[9] Laudable projects for broader global benefit from Western-based medical innovation, to combat disparities, and for a less homogenous scientific workforce flow together in the valuation of diverse bioinformation in medical projects.

I return here to the example of the NIH precision medicine initiative to draw out its implications for an anthropology of bioinformation in some more detail. This research program has indeed declared diversity one of its core tenets (cf. Sabatello and Appelbaum 2017). Established by former U.S. president Obama in 2015 as the Precision Medicine Initiative (PMI), the national program was to mark a turn from a one-size-fits-all medicine to a projected future individual-bioinformation-based, and thus truly *personalized*, genomic medicine model (White House 2015). This personalization in essence works through the biological stratification of patients (Day et al. 2017). PMI has only recently rebranded itself "All of Us," moving away from the individual and to the collective now interpellated in its name.

All of Us has produced countless creative advertisements claiming that diverse data are needed for the progress of America's diverse communities. For instance, the program website states that "[b]y joining the program, you can … [h]elp represent your community in important studies that may lead to new research findings, treatments, and cures" (NIH n.d.b). Beyond the promise of cures, diverse participation thus signals as a promise of *political* representation even when *genetic* representation is at stake (which is to say, increased sample size). The racial logic of how individuals come to represent genetic populations is embedded within the notion of scaling up bioinformation (Fullwiley 2014, M'charek 2000).

A promise of agency also signals in this notion of being the bodily owner of diversity, of the diverse data that one may choose to share or not for the sake of community. *All of Us* advertises precision medicine through a direct claim that America's ethnic, racial and sexual minorities should participate, that their diverse bodies require studies using their diverse data. Capturing this diverse bioinformation is the way to ensure future benefit for these groups: diverse data are valuable because they hold the key to future benefit in terms of drugs and new treatments. While I do not question the sincerity of this and other projects to improve the health of U.S. minorities, the recruitment of "minority data" has long been criticized as exploitative of already marked and racialized bodies that are tokenized for PR or, on a global level, become literal biocapital through their limited exposures to drugs (Epstein 2007, Sunder Rajan 2006, 2012, Cooper 2012).

In *All of Us*, community benefit and biocapital merge seamlessly as diverse bodies and their people are conjured as property owner-agents of their own biological information, which they should make available for the sake of their small-scale group. Living in the U.S., we already know what this community is, how much the term "community" itself signals as the Other of an enduring baseline norm of a straight/white/male individual. Indeed, images of Black, Brown, and non-normatively gendered bodies confirm on program posters and advertisements how diversity attaches to certain bodies (Ahmed 2012).

While still collecting bioinformation from individuals rather than social scientific data about collectives, *All of Us* has firmly shifted its public outreach to the idioms of diversity and community. The program website states:

> Diversity in a research program is important for several reasons. First, where we live, how we live, and our background can all affect our health. Second, many groups of people have been left out of research in the past. This means researchers know less about their health. By studying data from a diverse group of people, researchers can learn more about what

makes people sick or keeps them healthy. What researchers learn could lead to better treatment and disease prevention for all of us.

Explaining why diversity is important in medical research, the quote reveals tensions between public relations and the actual research centering on bioinformation—ethnographies of how people live, for instance, have yet to be funded in the program. But when the NIH cites "where we live, how we live, and our background" as motivating reasons for seeking diverse participants, it seems to indicate that diversity itself will facilitate a move toward a better, more personal kind of medicine.

In an ethnographic study of the implementation of precision medicine in a British cancer clinic, anthropologist Sophie Day et al. (2017) examined whether more personal medicine actually results from projects that rely on the biological stratification of patients. The authors found that, on the contrary, stratified medicine (under which they subsume personalized and precision medicine) entails ever-more nuanced divisions of labor that actually de-personalize medicine, because treatments are divided up into increasing numbers of visits to specialists whom patients often only see once. Clearly the meanings of precision and personalization in medicine remain contentious as even individually tailored care cannot achieve personal specificity in a biologically stratified sense (Tutton 2014). This problematic speaks to the tensions in bioinformation as specific yet universal, individual yet collective forms, where issues of commensurability, grouping, and representation constantly arise.

Day et al.'s study also evidences how much patient biological stratification fosters an ever-more diversified medical marketplace. Indeed anthropologist Sandra Lee has usefully framed precision medicine, created on the U.S. national stage with *All of Us* but echoed in countless iterations of new precision health centers across the country, as a neoliberal framework of medicine in which "disease and its management" are constructed "through economic rationalities" (2017: 35) within the paradigm of health as an industry and of diversity as a factor that aids economic development.

The workforce should represent diversity, the economy should be fueled by diversity, and bioinformation's value be directly tied to its provenance from diverse bodies. In the current moment, NIH researchers constantly claim to be doing community-oriented research. But in the end, their professional predicament requires biological information—it's the thing they'll do the science on—even if the community is interested in other interactions or tired of being sampled for bodily materials. In *All of Us*, participation in research is framed as an altruistic act, though projected benefit is now at the collective scale of community, where diversity speculatively signals as a more personal medicine for individuals from underserved racial groups.

In the above examples, bioinformation is most often called upon to represent ethnic and racial diversity. Historical reasons for choosing not to engage biomedical research are reframed as a problem of trust to be overcome, as advertisements promise marginalized communities benefits from a still pending genomic revolution in medicine. Contrasting starkly with CODIS' colorblind logic, diversity here lies within bioinformation from diverse bodies and animates projects imagined to undo historical wrongs through representation in scientific research, in a domain that is supposedly magic's rational opposite (cf. Jones 2018). And yet the idea of ethnic and racial diversity is itself grounded in American racial hierarchies. The described data practices do not illustrate this biopolitical order, but rather *reconstitute* it in the registers of diversity, community, personalization, and genetic race.

Consent across biomedical and forensic domains

Before concluding, I will briefly elaborate on a second key difference between value regimes of bioinformation across social domains. This difference relates to how data are derived, namely, from consenting research participants in medicine and from coerced subjects in forensics. Consent (specifically informed consent) is a key term in biomedicine that centers on the educational and voluntary decision-making process preceding the use of human bioinformation in research (Faden and Beauchamp 1986). In U.S. biomedicine, the value of diverse bioinformation is arguably entwined with consent as itself an articulation of a redemptive racial politics. Consent signals politically to indicate the progress of human subjects research from historical medical exploitation, where ethnic and racial minorities (among other disadvantaged groups) were not told about the extraction, use, or value of their bodily materials. Consent assures participants that institutions will protect their information (cf. Epstein 2008, Skloot 2010).

Legally, consent regulates and often defines future uses of bioinformation, even if it does so only vaguely. For instance, *All of Us* asks voluntary participants to consent to their data being deposited in a national biobank for at least a decade (Lee 2017). Consumer genomic companies ask customers to sign user agreements that include various forms of consent. Declared consent is also what allows for bioinformation to become biocapital when data travel across different contexts and are exchanged as commodities.[10]

In CODIS, bioinformation comes without consent, and often from the groups medical researchers consider the hardest to convince to participate in studies. The person from whom this kind of bioinformation derives does not give consent to the process of sampling, sequencing and storing of information, or for its potential future uses in research, because this person has either been convicted of a crime or is the subject of a felony arrest and thus has no choice but to contribute.[11] The American Civil Liberties Union (ACLU) recently argued— unsuccessfully—that taking DNA from felony arrestees constitutes an unreasonable search and violates informational privacy (Haskell v. Brown 2018). Other U.S. court rulings from the last several decades determine that the state's need for identifying felony arrestees takes precedence over expectations of privacy, voiding a need for consent. This debate has been led differently in Europe, where Portugal, for example, has instituted processes for legal subjects consenting to DNA samples being taken (Machado and Silva 2009). Further research is needed to examine whether consent could be a useful trajectory for forensic uses of bioinformation in the U.S., in particular as new molecular technologies are rapidly being developed and applied.[12]

The two contexts of biomedical research and legal identification are distinct, and yet bioinformation from both domains can end up traveling across the contexts of consumer genomics, biomedicine, and the legal system. For instance, CODIS data can also be used in "identification research" (FBI 2019). Whether the entities doing this research are public or private is not revealed, and there is no consistent regulation in the U.S. for who can access CODIS information (Rosen 2003). By contrast to biomedical research, where value derives from the fact that consent was given by a human subject to further uses of "their" data, CODIS bioinformation is not claimed as "theirs" by human subjects. Signaling back to the agential part of diversity as aspect of a value regime in which consenting human subjects own their bioinformation and can choose to share or not, we might ask if CODIS data, neither owned nor consented, *can* be diverse at all.

Conclusion

In what contexts is bioinformation made to matter as diverse, and thus valuable? Across biomedicine and forensics in the U.S., social advantages and disadvantages end up reflected in the make-up of genetic databases and in bioinformation itself. In medical research, bioinformation matters as diverse data, and is valuable when it represents diversity, which is already salient from mundane institutional and policy practices. Inequality is also what many biomedical researchers aim to fight through their scientific practice. Racial inequality in particular is the background against which the value of diverse bioinformation is proclaimed, when its *diversity* is why this information is said to be important in new medical projects that address diverse communities. In CODIS, inequality is excluded from mattering, relegated to meta-data, yet the bodies from whom samples are taken are those same bodies to whom diversity has attached itself in America. And yet, because of the structural racism of the U.S. criminal justice system, this database might be the most racially varied.

Anthropologist Noah Tamarkin argues in a recent paper on forensic uses of DNA in South Africa that "biomaterial is translated into searchable disembodied numbers" in a process he calls data capture, whereby a sample becomes data on the screen (2019: 4). In Tamarkin's analysis, this "capture" neutralizes the threat of the perpetrator, because genetics' scientific certainty in and of itself promises safety. In CODIS, there is an aim for racial neutrality, and calls for a universal database seek to overcome systemic biases. Through a lens of data capture, I suggest that in CODIS the very groups that have in the American imagination been perceived as a threat are captured as bioinformation without consent, in speculative logics of neutralizing risk. Meanwhile, biomedical data circulate in scientific and economic contexts that put a premium on capturing diversity.

In this chapter, I examined different kinds of bioinformation contexts in which geneticists and others make universalizing claims at different scales. Calls for the need for data- and researcher-diversity in biomedicine rely on logics of bodily difference, data ownership and economic growth. Here diversity signals as a marker of DNA itself, and as a sign of a more personalized medicine that takes seriously the needs of historically underrepresented and medically underserved American communities. By contrast, experts evoke universality as the potential solution to racial bias in CODIS. The first instance lacks diverse participants and is grounded in an ethical paradigm of consenting subjects in research, who voluntarily share their own bioinformation, itself considered to contain diversity. CODIS is grounded in colorblind ideas about junk DNA as meaningless bioinformation that does not reflect diversity. The second instance is rich in "diverse participants" and derives bioinformation for legal identification without the necessity for consent.

I thus explicated that data that lack diversity or "diverse data" are not stable entities. Diverse data emerge amidst Fortun's ongoing chiasmus, which is to say amidst tensions around bioinformation's very nature as specific yet universal, and as individual yet collective. Diversity is ascribed to data in contexts that sustain different epistemologies and politics, and the over- and under-representation of specific groups acquires different meanings and values across medicine and forensics. As data travel across regulatory regimes, CODIS and medical bioinformation may yet cross over from one to the next. In such "data journeys," philosopher Sabina Leonelli argues, "[a]ttributions of scientific value are … closely intertwined with attributions of political, economic, and affective value to data" (2016: 5). In sum, overlapping American racial politics of diversity and colorblindness, different epistemologies of DNA, and different calculations of biocapitalism operate in these competing universalisms of contemporary genetics.

Cases cited

Haskell *v.* Brown, 317 F.Supp.3d 1095 (2018)
    Maryland *v.* King, 569 U.S. 435 (2013)

Notes

1 My dissertation fieldwork examined the uses of genetic technologies across different social domains in California. In two postdoctoral positions in U.S. medical schools, I researched team diversity in medical innovation, and the emerging field of precision psychiatry.
2 The NIH defines this approach as the "precise delineation of the molecular, environmental, behavioral, and other factors that contribute to health and disease," with the goal of "more accurate diagnoses, more rational disease prevention strategies, better treatment selection, and the development of novel therapies." www.nih.gov/sites/default/files/research-training/initiatives/pmi/pmi-working-group-rep ort-20150917-2.pdf, accessed August 10, 2020. Day et al. define precision medicine as "sub-classification of disease … to enable recursive tailoring of treatment to individual response" (2017: 144).

3 https://allofus.nih.gov/about/diversity-and-inclusion, accessed July 28, 2020.
4 I follow Parry and Greenhough (2018) in their definition of bioinformation as "all information, no matter how constituted, arising from analyses of biological organisms and their behaviour ..." (2018: 8). My focus is on genetic data derived from human bodies.
5 As Helmreich (2008) outlines, a number of scholars have theorized the term *biocapital*, first among them Sarah Franklin (2003), who defines it in relation to bodily reproductive abilities. Kaushik Sunder Rajan's 2006 monograph tracks biocapital to mean both commodities from biological materials as well as new kinds of speculative capital. Kalindi Vora (2015) theorizes the gendered and affective labor of value creation in the production of biocapital from living human beings.
6 Writing about U.S. politics, I qualify this idea because the moment at stake has significantly changed since 2016. Diversity as a value still, arguably more than ever, connotes liberal politics and neoliberal policy. Yet the Trump administration has engendered few celebrations of diversity, faux or not. The Trump moment was marked by an overt return to white supremacist heteropatriarchy, where all deviations from the straight/white/male/American norm are disparaged.
7 Sociologist Eduardo Bonilla-Silva defines colorblindness as a reactionary ideology that arose during the U.S. civil rights movement. He identifies four paradigms, "abstract liberalism, naturalization, cultural racism, and minimization of racism" (2006: 54), as the main tenets of this enduring form, in which, he argues, racism prevails even though most claim that they are not racist.
8 Genome-wide association studies (GWAS) scan genomes to find associations (not necessarily causative) between genetic variants and particular traits.
9 A similar discussion takes place in AI and algorithmic systems, where a focus on eliminating bias by ensuring that data are *good* can obscure other political issues, see Courtland 2018.
10 For instance, consumer genetic testing companies sell bioinformation, if consumers have consented, to biomedical research companies in bundles. Data that were initially meaningful to consumers, and what they purchased in the form of a test, become biocapital only in a second economic exchange in which such companies' actual profit is made, see www.vice.com/en_us/article/xwkaz3/23andme-sold-access-to-your-dna-library-to-big-pharma-but-you-can-opt-out, accessed March 19, 2020.
11 In a study of algorithmic personalization, Lury and Day define such unknowing participation as "participative," concluding that medical personalization not only represents a "precise form of individuation [but] also constrains who and how we can be" (2019: 19–21).
12 One such technology, used in criminal investigations, is forensic DNA profiling, which relies on a strongly racialized logic (M'charek et al. 2020, Murphy 2018, Vailly, in press).

References

Abu El-Haj, Nadia (2012) *The Genealogical Science: The Search for Jewish Origins and the. Politics of Epistemology*. Chicago: University of Chicago Press.
Ahmed, Sara (2007) The Language of Diversity. *Ethnic and Racial Studies*, 30 (2): 235–256.
Ahmed, Sara (2012) *On Being Included: Racism and Diversity in Institutional Life*. Durham, NC: Duke University Press.
Arciniega, Luzilda Carrillo (2019) What Does Diversity and Inclusion Mean? *Anthropology News* website, February 7. doi:10.1111/AN.1090.
Aronson, Jay (2007) *Genetic Witness: Science, Law, and Controversy in the Making of DNA. Profiling*. New Brunswick: Rutgers University Press.

Bonilla-Silva, Eduardo (2006) *Racism Without Racists: Color-blind Racism and the Persistence of Racial Inequality in the United States*. Lanham: Rowman & Littlefield.

Bowker, Geoffrey (2005) *Memory Practices in the Sciences*. Cambridge: MIT Press.

Brown, Wendy (2015) *Undoing the Demos: Neoliberalism's Stealth Revolution*. New York: Zone Books.

Cho, Mildred & Pamela Sankar (2004) *Forensic Genetics and Ethical, Legal and Social Implications Beyond the Clinic*. Nature Genetics, 36 (11 Suppl): S8–12.

Cooper, Melinda (2012) Pharmakologie im Zeitalter des verteilten Experiments. In Susanne Lettow(ed.), *Bioökonomie. Die Lebenswissenschaften und die Bewirtschaftung der Körper*. Bielefeld: transcript, pp. 109–132.

Courtland, Rachel (2018) The Bias Detectives. *Nature*, 558: 357–360.

Day, Sophie, R. Charles Coombes, Louise McGrath-Lone, Claudia Schoenborn & Helen Ward (2017) Stratified, Precision or Personalised Medicine? Cancer Services in the 'Real World' of a London Hospital. *Sociology of Health and Illness*, 39 (1): 143–158.

Epstein, Steven (2007) *Inclusion: The Politics of Difference in Medical Research*. Chicago: University of Chicago Press.

Epstein, Steven (2008) The Rise of "Recruitmentology": Clinical Research, Racial Knowledge, and the Politics of Inclusion and Difference. *Social Studies of Science*, 38 (5): 801–832.

Faden, Ruth & Tom Beauchamp (1986) *A History and Theory of Informed Consent*. Oxford: Oxford University Press.

Fausto-Sterling, Anne (1992) *Myths of Gender: Biological Theories about Women and Men*. New York: Basic Books.

Federal Bureau of Investigation (2019) National DNA Index System (NDIS) Operational Procedures Manual. www.fbi.gov/file-repository/ndis-operational-procedures-manual.pdf/view, accessed March 19, 2020.

Ferguson, Roderick (2012) *The Reorder of Things: The University and its Pedagogies of Minority Difference*. Minneapolis: University of Minnesota Press.

Fortun, Mike (2008) *Promising Genomics: Iceland and deCODE Genetics in a World of Speculation*. Berkeley: University of California Press.

Franklin, Sarah (2003) Ethical Biocapital: New Strategies of Cell Culture. In Sarah Franklin & Margaret Lock (eds.), *Remaking Life and Death: Toward an Anthropology of the Biosciences*. Santa Fe: SAR Press, pp. 97–127.

Fullwiley, Duana (2007) The Molecularization of Race: Institutionalizing Human Difference in Pharmacogenetics Practice. *Science as Culture*, 16 (1): 1–30.

Fullwiley, Duana (2014) The "Contemporary Synthesis": When Politically Inclusive Genomic Science Relies on Biological Notions of Race. *Isis*, 105 (4): 803–814.

Garrison, Nanibaa' (2013) Genomic Justice for Native Americans: Impact of the Havasupai Case on Genetic Research. *Science, Technology, & Human Values*, 38 (2): 201–223.

Hammonds, Evelynn & Banu Subramaniam (2003) A Conversation on Feminist Science Studies . *Signs: Journal of Women in Culture and Society*, 28 (3): 923–944.

Haraway, Donna (1988) Situated Knowledges: The Science Question in Feminism and the Privilege of Partial Perspective. *Feminist Studies*, 14 (3): 575–599.

Haraway, Donna (1989) *Primate Visions: Gender, Race, and Nature in the World of Modern. Science*. New York: Routledge.

Harding, Sandra (1986) *The Science Question in Feminism*. Ithaca: Cornell University Press.

Harrington, Mark (2008) AIDS Activists and People with Aids: A Movement to Revolutionize Research and for Universal Access to Treatment. In Kavita Philip & Beatriz Da Costa (eds.), *Tactical Biopolitics: Art, Activism, Technoscience*. Cambridge: MIT Press, pp. 323–340.

Harry, Debra & Jonathan Marks (1999) Human Population Genetics Versus the HGDP. *Politics and the Life Sciences*, 18 (2): 303–305.

Hazel, James, Ellen Clayton, BradleyMalin, ChristopherSlobogin (2018) Is It Time For A Universal Genetic Database? *Science*, 362 (6417): 898–900.

Helmreich, Stefan (2008) Species of Biocapital. *Science as Culture*, 17 (4): 463–478.

Hewlett, Silvia, Melinda Marshall & Laura Sherbin (2013) How Diversity Can Drive Innovation. *Harvard Business Review*, https://hbr.org/2013/12/how-diversity-can-drive-innovation, accessed July 31, 2020.

Jabloner, Anna (2019) A Tale of Two Molecular Californias. *Science as Culture* 28 (1): 1–24.

Jabloner, Anna & Sandra S.Lee (2020) Who is the Right Fit? Doing Diversity in Translational Research. *Catalyst: Feminism, Theory, Technoscience*, 6 (1): 1–24.

Jones, Graham (2018) *Magic's Reason: An Anthropology of Analogy*. Chicago: University of Chicago Press.

Kaye, David & Michael Smith (2004) DNA Databases for Law Enforcement: The Coverage Question and the Case for a Population-Wide Database. In David Lazer (ed.), *DNA and the Criminal Justice System: The Technology of Justice*. Cambridge: MIT Press.

Keller, Evelyn Fox (1982) Feminism and Science. *Signs: Journal of Women in Culture and Society*, 7 (3): 589–602.

Lee, Sandra (2008) Racial Realism and the Discourse of Responsibility for Health Disparities in a Genomic Age. In Sandra Lee, Barbara Koenig & Sarah Richardson (eds.), *Revisiting Race in a Genomic Age*. New Brunswick: Rutgers University Press, pp. 342–358.

Lee, Sandra (2017) Consuming DNA: The Good Citizen in the Age of Precision Medicine. *Annual Review of Anthropology*, 46: 33–48.

Lee, Sandra, Stephanie Fullerton, Aliya Saperstein & Janet Shim (2019) Ethics of Inclusion: Cultivate Trust in Precision Medicine. *Science*, 364 (6444): 941–942.

Leonelli, Sabina (2016) *Data-Centric Biology: A Philosophical Study*. Chicago: University of Chicago Press.

Lombardo, Paul (2008) *Three Generations, No Imbeciles: Eugenics, the Supreme Court, and Buck v. Bell*. Baltimore: Johns Hopkins University Press.

Lowe, Lisa (1996) *Immigrant Acts: On Asian American Cultural Politics*. Durham, NC: Duke University Press.

Lury, Celia & Sophie Day (2019) Algorithmic Personalization as a Mode of Individuation. *Theory, Culture & Society*, 36 (2): 17–37.

M'charek, Amade (2000) Technologies of Population: Forensic DNA Testing Practices and the Making of Differences and Similarities. *Configurations*, 8 (1): 121–159.

M'charek, Amade, Victor Toom and Lisette Jong (2020) The Trouble with Race in Forensic Identification. *Science, Technology, & Human Values*, 45 (5): 804–828.

Machado, Helena & Susana Silva (2009) Informed Consent in Forensic DNA Databases: Volunteering, Constructions of Risk and Identity Categorization. *BioSocieties*, 4: 335–348.

Murphy, Erin (2015) *Inside the Cell: The Dark Side of Forensic DNA*. New York: Nation Press.

Murphy, Erin (2018) Forensic DNA Typing. *Annual Review of Criminology*, 1: 497–515.

Murphy, Erin & Jun Tong (2019) The Racial Composition of Forensic DNA Databases. NYU School of Law, *Public Law Research* 19–54, http://dx.doi.org/10.2139/ssrn.3477974, accessed August 2, 2020.

National Institutes of Health (n.d.a) https://grants.nih.gov/policy/inclusion/women-and-.minorities.htm, accessed March 17, 2020.

National Institutes of Health (n.d.b) https://allofus.nih.gov/about/diversity-and-inclusion, accessed March 18, 2020.

Need, Anna & David Goldstein (2009) Next Generation Disparities in Human Genomics: Concerns and Remedies. *Trends in Genetics*, 25 (11): 489–494.

Ordover, Nancy (2003) *American Eugenics: Race, Queer Anatomy, and the Science of Nationalism*. Minneapolis: University of Minnesota Press.

Parry, Bronwyn & Beth Greenhough (2018) *Bioinformation*. Cambridge: Polity Press.

Popejoy, Alice & Malia Fullerton (2016) Genomics is Failing on Diversity. *Nature*, 538: 161–164.

Pratt, Mary Louise (1992) *Imperial Eyes: Travel Writing and Transculturation*. New York: Routledge.

Reardon, Jenny (2005) *Race to the Finish: Identity and Governance in an Age of Genomics*. Princeton: Princeton University Press.

Reardon, Jenny (2017) *The Postgenomic Condition: Ethics, Justice, and Knowledge After the Genome*. Chicago: University of Chicago Press.

Reardon, Jenny & Kim TallBear (2012) "Your DNA Is Our History": Genomics, Anthropology, and the Construction of Whiteness as Property. *Current Anthropology*, 53 (S5): S233–S245.

Regalado, Antonio (2018) White-People-Only DNA Tests Show How Unequal Science has Become. *MIT Technology Review*, www.technologyreview.com/s/612322/white-peop le-only-dna-tests-show-how-unequal-science-has-become/, accessed March 18, 2020.

Reverby, Susan (2009) *Examining Tuskegee: The Infamous Syphilis Study and Its Legacy*. Chapel Hill: University of North Carolina Press.

Rosa, Jonathan & Yarimar Bonilla (2017) Deprovincializing Trump, Decolonizing Diversity, and Unsettling Anthropology. *American Ethnologist*, 44 (2): 201–208.

Rosen, Christine (2003) Liberty, Privacy, and DNA Databases. *The New Atlantis*, 1: 37–52.

Sabatello, Maya & Paul Appelbaum (2017) Precision Medicine Nation. *Hastings Center Report*, 47 (4): 19–29.

Shankar, Shalini (2015) *Advertising Diversity: Ad Agencies and the Creation of Asian American Advertising*. Durham, NC: Duke University Press.

Sirugo, Giorgio, Scott Williams & Sarah Tishkoff (2019) The Missing Diversity in Human Genetic Studies. *Cell*, 177 (1): 26–31.

Skloot, Rebecca (2010) *The Immortal Life of Henrietta Lacks*. New York: Random House.

Slaughter, Sheila & Gary Rhoades (2004) *Academic Capitalism and the New Economy: Markets, State, and Higher Education*. Baltimore: Johns Hopkins University Press.

Stern, Alexandra M. (2016) *Eugenic Nation: Faults and Frontiers of Better Breeding in Modern America*. Berkeley: University of California Press, 2nd edition.

Stevenson, Anne & Lukoye Atwoli (2019) Moving Away From "White People Only" DNA Tests: African Project Seeks Thousands For Mental Health Genetics. *WBUR Boston's NPR News Station*, www.wbur.org/commonhealth/2019/08/13/solving-di versity-problem-schizophhrenia-genetics-research, accessed March 18, 2020.

Subramaniam, Banu (2000) Snow Brown and the Seven Detergents: A Metanarrative on Science and the Scientific Method. *Feminist Studies*, 28 (1/2): 296–304.

Subramaniam, Banu (2009) Moored Metamorphoses: A Retrospective Essay on Feminist Science Studies. *Signs: Journal of Women in Culture and Society*, 34 (4): 951–980.

Subramaniam, Banu (2014) *Ghost Stories for Darwin: The Science of Variation and the Politics of Diversity*. Champaign, IL: University of Illinois Press.

Sunder Rajan, Kaushik (2006) *Biocapital: The Constitution of Postgenomic Life*. Durham, NC: Duke University Press.

Sunder Rajan, Kaushik (2012) Introduction: The Capitalization of Life and the Liveliness of Capital. In Kaushik Sunder Rajan (ed.), *Lively Capital: Biotechnologies, Ethics, and Governance in Global Markets*. Durham, NC: Duke University Press, pp. 1–41.

Tamarkin, Noah (2019) Forensics and Fortification in South African Self-Captivity. *History and Anthropology*, 30 (5): 1–6.

Tutton, Richard (2014) *Genomics and the Reimagining of Personalized Medicine*. Farnham: Ashgate.

Ulrich, Tom (2018) In Psychiatric Genetics, A Push to Think Globally, Act Locally. *Broad Institute*, www.broadinstitute.org/news/psychiatric-genetics-push-think-globally-act-locally, accessed March 18, 2020.

Urciuoli, Bonnie (2011) Neoliberal Education: Preparing the Student for the New Workplace. In Carol Greenhouse (ed.), *Ethnographies of Neoliberalism*. Philadelphia: University of Pennsylvania Press, pp. 162–176.

Urciuoli, Bonnie (2016) The Compromised Pragmatics of Diversity. *Language & Communication*, 51: 30–39.

Vailly, Joëlle. In press. Appearance and Origin: The Depoliticization of Genetic Privacy in France. *Current Anthropology*.

Vora, Kalindi (2015) *Life Support: Biocapital and the New History of Outsourced Labor*. Minneapolis: University of Minnesota Press.

Washington, Harriet (2006) *Medical Apartheid: The Dark History of Medical Exploitation on Black Americans from Colonial Times to the Present*. New York: Harlem Moon.

Weiss, Hadas (2015) Capitalist Normativity: Value and Values. *Anthropological Theory*, 15 (2): 239–253.

White House. 2015. FACT SHEET: President Obama's Precision Medicine Initiative. Off. Press Secr. Press Release, Jan. 30. https://obamawhitehouse.archives.gov/the-press-office/2015/01/30/fact-sheet-president-obama-s-precision-medicine-initiative, accessed March 18, 2020.

Williamson, Robert & Roni Dunkel (2002) DNA Testing For All. *Nature*, 418 (6898): 585–586.

8 Global e-waste epidemiology and emerging politics of bioinformatic extraction

Peter C. Little

Introduction

Recent developments in global health studies of electronic waste (e-waste) are a topic at the anthropological crossroads of science practice and expertise, labor, and toxic bioinformatics. As accounts of e-waste disasters in Asia and Africa began to emerge in the early 2000s, international science and advocacy interests and interventions began to emerge, leading to a variety of solutions-based waste management projects and, more recently, environmental health studies. Beginning in 2018, the U.S. National Institutes of Environmental Health Sciences (NIEHS) even began to offer webinars to report on epidemiological evidence generated from case studies in China and Ghana, and the number of scientific reports exploring e-waste and health relations is on the upswing. These emerging environmental health science projects are producing e-waste biodata and measurable bio-evidence that might rightfully inform global e-waste management and policy, but there exist particular knowledge extraction and epistemic politics amidst these epidemiological studies that call for careful environmental health critique and critical bioinformatic reflection.

To navigate these underlying biopolitics of global e-waste health studies, here I use the term *bioinformatic extraction* to signal the testing of bodies and extraction of biodata—like blood, semen, and maternal urine—to bolster e-wastes' epidemiological evidence. I draw on and compare e-waste health science studies in Ghana and China,[1] revealing how these sites of bioinformatic extraction guide contemporary e-waste epidemiological knowledge.

Engaging in bioinformatic critique involves, for me, tracking and attending to the global travels of e-waste health science and their emergent biopolitical consequences. In light of this, the chapter is inspired by several interlinked questions that might open e-waste anthropologies up to multi-sited and comparative bioinformatic critique: What communities and bodies are targeted for global e-waste health studies? What bioinformation is extracted from these situated economies, these labor cohorts and environments, and for what purpose? What biosocial politics emerge and matter in these zones of bioinformatic extraction? What metrics matter and make sense to e-waste epidemiological science? Finally, what is the value of bioinformatic critique amidst emerging geographies of e-waste health

DOI: 10.4324/9780367810030-8

science and expertise? These questions, it will be argued, help guide a critical anthropology of e-waste bioinformation and can ultimately help advance theories of e-waste biopolitics.

Bioinformatic critique and the emergence of e-waste epidemiology

Bioinformation can be understood as a relational or synthesis concept that helps us rethink the body and embodiment amidst, among many other things, a broader anthropology of science, data, and labor. This anthropology involves a certain political ecological awareness of the plethora of interlinked exposures and forms of embodiment and bodily entanglement flourishing on our "damaged planet" (Tsing et al. 2017), a planet overwhelmed with contaminants of capitalism, from plastic water bottles to toxic electronic and electrical equipment discard. This situation has sparked a parallel interest in the emergence of certain and variegated "biosocial matters" (Meloni, Williams, and Martin 2016), one of which is biosocial mattering linked to the worlding of wastes like e-waste (Lepawsky 2018). E-waste is currently an expanding trade sector of the global economy, especially an emerging circular economy increasingly based on reuse materials and reuse extraction labor. The UN, for example, has claimed this sector generated nearly $US18 billion in 2017, with growth on the upswing according to a recent World Bank report (Kaza et al. 2018). The bodies that handle this e-waste, from Africa to Asia, embody and signify a pained Earth marked by a confluence of forces, big and small, from large scale mining of metals and minerals, massive agribusiness land grabs and landscape transformations, countless cases of industrial pollution and toxic spillage, to obsolescence-based production logics and the manufacture of futuristic gadget desires that seem to only grow larger and more intrusive each day. This global situation has tilted much conversation in anthropology towards environmental health research (Singer 2016) and a more general anthropology of toxics, pollution, and waste that attends to ideas and practices that blur (and breakdown) situated boundaries and bodies, economies and ecologies, exposures and bioaccumulations, science and uncertainty, new toxic risks and waste materialities (Little 2019, Lock 2017, Reno 2015, 2016, Tousignant 2018, Isenhour and Reno 2019). This chapter draws on these anthropological perspectives to explore how toxic e-waste is embodied, what this particular bioinformation tells us about our hectic electronics age, and how e-waste workers enrolled in global e-waste epidemiology studies become sources and subjects of bioinformation amidst the broader "hegemony of biometrics" (Walkover 2016).

As an environmental and medical anthropologist, I take e-waste risk and toxicity to be *known* through working bodies and active subjectivities (Reno 2016). E-waste labor takes a more central position in the toxic analysis of e-waste recycling, and because of this, the boundary between labor and toxic substances collapses. Data on labor "activities" and epidemiological data are being integrated and resituated, exposing how global e-waste epistemics are conditioned and actively made by practices of bioinformatic extraction that rely on "situated biologies" (Lock 2017) and bodies at work. Analyzing the epistemic promises and biopolitical consequences of this situated extraction, it will be argued, is integral to building an anthropology of e-waste attuned to an emergent global environmental health science practice engaged in the production of more, not fewer, metrics (Adams 2016). Currently, e-waste tracking[2] and ill-health tracking are being synthesized to build a more evidence-based global health perspective on e-waste and its urban environmental health dimensions. This is following the same trajectory of other global health programs and trends. As Tichenor (2016: 105, emphasis added) points out, "Current forms of global health funding require *tightly regulated evidence*, proving to the international aid community that their aid has real impact on the local situation, to justify supporting health programs in the Global South." Moreover, this biometric evidence is largely needed or made governable because e-waste disaster zones are approached as uncertain urban environmental hazards (Zeidman 2015).

But, while the emergence of health science expertise in e-waste studies appears new, the examination and interrogation of the electronics industry and ill-health relationship is nothing new. Since the publication of *Challenging the Chip: Labor Rights and Environmental Justice in the Global Electronics Industry* (Smith, Sonnenfeld and Pellow 2006), knowledge of the risks and toxicity of electronics industry work and e-waste recycling work has been well understood. The book was the first to highlight the global dimensions and complexities of toxic situations in which workers in the electronics industry and those working in e-waste recycling labor find themselves (LaDou 2006). Since the mid-2000s, environmental epidemiologists have emphasized the various risks associated with e-waste worker exposures to a variety of toxic metals. Toxic substances like lead (Pb), mercury (Hg), cadmium (Cd), hexavalent chromium (Cr), brominated flame retardants (PBDEs), as well as polychlorinated biphenyls (PCBs) and persistent organic pollutants (POPs), have been the primary toxics of concern. Multiple science agencies, from the U.S. Centers for Disease Control and Prevention (CDC), the National Institutes for Environmental Health Sciences (NIEHS), the Environmental Protection Agency (EPA), to the World Health Organization (WHO), recognize the public health risks of exposures to these highly toxic substances.

The direct biological impacts of lead exposures have seen perhaps the greatest attention in e-waste studies. Lead (Pb) is a widely known developmental neurotoxicant and has become a central focus of e-waste health studies, as well as a focus of recent anthropological research (Renfrew 2018). For example, it has been reported that old televisions (CRT or cathode ray tube) and old CRT computer monitors contain significant levels of lead (U.S. EPA 2007, 2008). Research on children's exposures to lead in e-waste recycling sites has shown that mean blood levels in children between one and six years old can be extremely high (15 μg/dL) (Huo et al. 2007, Zheng et al. 2008), a biometric that raises serious concerns about direct impacts on neurodevelopment, like cognitive function. Mercury (Hg) is yet another toxic substance that can bioaccumulate in e-waste workers. Mercury is used in laptop monitors, cell phones, cold cathode fluorescent lamps, and printed circuit boards (e.g., switches and relays), some e-waste recycling methods can release vaporized mercury into the environment, posing yet another concerning health risk (U.S. EPA 2007). Additionally, some note that the very acceleration of high-tech production has directly contributed to a truly global health threat that won't slow down until there is a significant reduction in e-production and e-waste trends (Aich et al. 2020).

The fact is considerable evidence of the health risks of e-waste recovery and recycling labor exists. Yet, according to public health experts, while e-waste health epistemics are on the upswing,

> serious data gaps exist in the quantification of exposures and health effects. In communities where informal e-waste recycling occurs, biomonitoring of exposures, especially in vulnerable pregnant women and young children, provide critical information for epidemiologic investigations, environmental policy making, and informed plans for intervention. Studies that use sensitive neurodevelopmental end points are particularly important in this complex exposure. Other potential toxicities in humans—for example, cancer, respiratory diseases, reproductive functions, and renal effects—should also be examined.
>
> (Chen et al. 2011: 436)

E-wastes' bioinformatic hubs

What follows is a brief sketch of two e-waste study areas, two e-waste biodata research hubs, that have dominated e-waste health science studies to date. I focus in particular on epidemiological research on e-waste in China and West Africa, two regions of the world with meaningful differences—cultural, political, economic, etc.—and even differences in the way in which science is governed and how e-waste epidemiology is conducted. I begin with e-waste health science in China, where a colossal electronics industry has been a key driver of the country's overall economic growth.

E-waste epidemiology in China

China has long been a geopolitical epicenter of electronics production and consumption (Leong and Pandita 2006). Guiyu, China has since the 1990s been a focal point of e-waste recycling studies (M. H. Wong et al. 2007, C. S. Wong et al. 2007a, C. S. Wong et al. 2007b). A rice-growing town located in Shantou, Guangdon Province, Southeast China, Guiyu is traditionally known as the "treasure city" and over the last two decades has become one of the world's largest e-waste recycling sites. Roughly 15,000 metric tons of e-waste enter Guiyu each day, becoming a robust recycling hub for China's larger circular economy (Schulz and Lora-Wainwright 2019). Currently, more than eighty percent of families in Guiyu engage in e-waste recycling labor, many of whom don't wear protective equipment to protect themselves from toxic heavy metal exposures (Kim et al. 2019). E-waste recycling sites in Guiyu are mainly located within or adjacent to residential areas and are completely surrounded by farms, especially in the southern part of Guiyu. Additionally, while Guiyu's population is around 150,000, the migrant laborer population hovers around 100,000 people (M. H. Wong et al. 2007, C. S. Wong et al. 2007a). Recent reports note that Guiyu's emphasis on e-waste recycling is disappearing as a result of e-waste recycling hubs moving to urban zones in mainland China, Hong Kong, and other cities in southeast Asia (Standaert 2015). But, regardless of where e-waste recycling businesses relocate, bodily contamination persists.

Toxic metal concentrations have been detected in the blood and maternal urine of Guiyu workers, with some recent studies emphasizing the risks of these toxic biodata in pregnant women and neonates (Kim et al. 2019). In 2011, scientists at the Shantou University Medical College and the University of Cincinnati spearheaded an environmental epidemiology project called the e-waste Recycling Exposures and Community Health (e-REACH) study, which analyzed the maternal blood, core blood, and maternal urine of 634 pregnant women living in Guiyu and Haojiang (control site). The researchers found that pregnant women in Guiyu were at increased risk for heavy metal exposures, especially lead, cadmium, nickel, and manganese. The public health scientists doing this research note that their study is unique amidst other e-waste epidemiological studies:

> This is one of the first studies to comprehensively capture maternal and neonate biological samples from a community with intensive informal e-waste recycling and test for metals exposure. In addition, the simultaneous collection of environmental and biological samples confirmed the widespread contamination of heavy metals in Guiyu.
>
> (Kim et al. 2019: 412)

The comprehensive "capture" of new biodata on e-waste risk in Guiyu is, in many ways, telling of the ongoing-ness of e-waste epidemiology in China, the golden epicenter of electronics production, consumption, and discard.

Given the expansive nature of e-waste work in China in recent years, it is important to note that the study of e-waste health risk in China expands beyond the Guiyu case study. Residents of Taizhou, in Zhejiang Province in East China, also engage in e-waste recycling and experience elevated levels of heavy metals and other pollutants, like dioxin (Chan et al. 2007, Tang et al. 2010). Researchers also emphasize how fetal exposure to many of these toxins, in Guiyu and other informal e-waste recycling communities around the world, continue beyond birth and postnatal exposures and are associated with health problems later in life. Another study of e-waste recycling around Tiajin, in northern China, takes a toxicogenomic approach that finds an association between DNA damage (including chromosomal aberrations) and elevated heavy metal concentrations in the peripheral blood and semen of e-waste recyclers

(Wang et al. 2018). The study focused on a cohort of 146 male residents of Tiajin who were directly engaged in recycling e-waste, especially recycling of discarded computers, TV sets and cathode ray tube (CRTs) monitors. Researchers found a "strong correlation between instability of the genome and duration of exposure to contaminants from handling and processing e-waste," adding that "[d]espite the benefits to society from recycling e-waste ... workers should be monitored to ensure that they are not exposed to toxicants released during recycling activities" (Wang et al. 2018: 78). These scientists admit that despite the evidence of certain toxic biodata, uncertainty endures. The toxicgenomic impacts of e-waste recycling labor, they contend, could be the result of heavy metal exposures leading to genetic mutations and disturbances, including biomarkers such as oxidative damage to nucleobases, induction of membrane lipid peroxidation, DNA methylation, and dysfunction of DNA repair.

While the health studies mentioned above draw on human biomarkers and biodata to showcase the toxicity of e-waste, according to Schulz (2015), much of the focus on e-waste in China has been couched within the terms of waste management, rather than within the framework of public health toxicology.

> The scientific community in China ... played, and continues to play, an important role in the conception and handling of used e-appliances as a form of waste. The discourse on "e-waste" or "waste electrical and electronic equipment" (WEEE), which now dominates in academic and policy-making circles around the world, implies that solutions are to be sought in the field of "waste management" (feiwu guanli 废物管理). In China, research on discarded e-appliances therefore focuses on material recovery and pollution control, and is mostly conducted by specialists of environmental sciences.

The recent growth in attention to e-waste epidemiology in China might very well lead to more interdisciplinary studies that blend different scientific fields. But, whether or not this becomes the new trend in e-waste studies in China, to really understand current and near-future e-waste science and technology matters in China calls for centering attention on the force of state control over science and technology in post-Mao China writ large. As Greenhalgh and Zhang (2020: 3) recently note,

> The post-Mao years have brought the rapid development not only of science, but also of *scientism,* the belief in science as a panacea for all the nation's ills ... Post-Mao China has been home to a veritable state-sponsored religion of S&T [science and technology] marked by a widespread faith in the power of modern science and technology to solve the problems that other approaches have failed to solve.

Another interesting thread in China's e-waste science narrative is the fact that many of the current e-waste epidemiological studies there are placing a stronger emphasis on reproductive environmental health and epigenetics, which is involving the mass enrollment of women in research projects. As will be discussed later, this is not the case for emerging e-waste epidemiology in Ghana. One likely reason for this is that in Guiyu, as well as other e-waste recycling sites in China, women engage directly in e-waste recycling labor. This is not the case in Ghana, where women support men working in e-waste recycling by selling food, water, and other goods.[3] The gendering of e-waste science studies, therefore, varies between China and Ghana, and exposes some of the divergent approaches taken based on variable socio-cultural context. So, while a bioinformatic critique of e-waste in China needs to account for larger state controls of science in general and public health toxicology in particular, the situation in Ghana is different, even if the experience of exposures for these informal e-waste laborers is shared across these radically different cultural, economic, and political spaces.

E-waste epidemiology in Ghana

Across Africa, there has been a steady growth in e-waste health studies (Orisakwe et al. 2019). Among the countries in focus, including Nigeria, Kenya, Tanzania, and Morocco, Ghana has for more than two decades been at the epicenter of Africa's e-waste narrative. Unlike e-waste health studies in China, a country that is a major node of global electronics production and discard, Ghana's story is largely one of e-waste recycling and toxic e-scrap economics in the Global South. The e-waste recycling site in Ghana that has seen the greatest focus is Agbogbloshie, an urban scrap metal market in the capital city of Accra. Beginning in the early 1990s, in an environment of rapid growth in informal sector activities around the broader Agbogbloshie area, e-waste processing quickly became a focus of the Agbogbloshie scrap market, which began as a prominent vehicle repair zone in Accra (Grant and Oteng-Ababio 2012, Davis, Akese and Garb 2018). Agbogbloshie was the focus of the first environmental health science projects focusing on e-waste labor in Ghana (Caravanos et al. 2011). This research was conducted by partnership between epidemiologists from Hunter College (New York) and Ghana Health Services, Ghana's primary public health agency. This early cohort study involved a small sample size (n = 5) and focused on urine samples to determine worker exposures to toxic substances (Feldt et al. 2014). Later studies that used larger sample sizes and better methods to link actual exposures and measured health outcomes found that e-waste sorters had high blood lead levels and those burning electronic cables to extract copper had high blood levels of copper and zinc (Srigboh et al. 2016).

Other studies in Agbogbloshie have shown strong associations between noise pollution—scrap metal markets are loud work environments—and increased heart rates among e-waste workers (Burns et al. 2016, Burns et al. 2019). This focus on measuring noise pollution is novel e-waste bioinformation and actually follows the recommendation of the WHO, which has defined noise exposure limits and is increasingly focusing on the environmental health burdens of occupational noise.[4] In addition to using a worker survey to capture "subjectively" perceived exposures to agitating noise while working, these scientists also "objectively" measured personal noise exposures in the Agbogbloshie scrap market by equipping workers with personal noise dosimeters (ER-200Ds). According to these researchers, extracting what I am calling e-waste bioinformation, is also involves measuring the noise pollution of the actual e-waste occupational soundscape:

> The dosimetry data were intended to compliment the subjective data by providing quantified estimates of noise exposure levels overall, as well as during particular work and non-work activities. These data were also collected to allow for assessment of the potential relationship between objective noise levels and injury risk.
>
> (Burns et al. 2019: 3)

Like other studies, these environmental health scientists note that future studies of e-waste need to be based on larger sample sizes and take a longitudinal approach since most risk characterizations for e-waste labor—locally and globally—involve informal labor practices that complicate *controlled* epidemiological studies of, for instance, migrant e-waste laborers, which make up the majority of those working in this sector in Ghana (Little 2019, Grant 2016).

West Africa has recently become the site of a newly emerging global e-waste health project called the West Africa GEOHealth Hub. This GEOHealth— short for Global Environmental and Occupational Health—project involves a partnership between the University and Ghana and the University of Michigan. The project is supported by an ongoing grant from the Forgarty International Center, a center of the U.S. National Institutes of Health. The Fogarty Center is a primary research division that supports global health science, especially in poor developing countries, and is among the few organizations to support environmental and occupational health research training in low and middle-income countries. The GEOHealth projects uses a "hub" model, creating regional centers for research and training and transnational science partnerships. "The hubs together form a network intended to serve as a platform to coordinate activities and provide a credible source for state-of-the-art knowledge on environmental and occupational health" (Fogarty Center website). According to the Fogarty Director Dr. Roger I. Glass, "The research hubs are designed to develop a critical mass of scientists who understand how the environment triggers disease, identify effective interventions and spur policy changes to improve health."

As a global health coordination strategy, GEOHealth projects[5] allow research scientists to keep transnational networks alive, especially research partnerships between the Global North and South. GEOHealth hubs are meant to foster science sustainability and build science capacity in places of significant resource, research, and training scarcity. According to the former director of the National Institutes of "Environmental and occupational health problems cross national boundaries, so research and training efforts to understand these problems through our GEOHealth hubs serves not only those affected locally, but all people suffering related issues," said NIEHS Director Dr. Linda Birnbaum. "Working with our partners to create sustainable research and training hubs in underserved countries benefits everyone." In many ways, these health science hubs—*bioinformatic hubs*—aim to regionalize and stabilize e-waste health epistemics while at the same time developing scientific partnerships and projects to foster "toxicological capacity" (Tousignant 2018), an approach to environmental health science that recognizes the technoscience and infrastructural demands of quality or scientifically-sound public health toxicology. Furthermore, and to increase networks of technoscientific expertise, these GEOHealth hubs are explicitly not designed to be isolated spaces of scientific knowledge but rather sites of data and knowledge that are institutionally—and epistemically—linked to U.S.-based academic and government research institutions and interests. Each of these research hubs addresses health risks prioritized in the respective region, and the research topics can range from outdoor and household air pollution, pesticide exposures, environmental contamination, climate change, and, for our purposes here, electronic waste.

To be fair, the recent e-waste turn for GEOHealth is in-part informed by a certain neoliberal political economy of science, a complex system of capital investments that ultimately help leverage scientific research support for GEOHealth projects and many other global health science projects around the globe. According to Sarah Felknor, research scientist with the National Institutes of Occupational Safety and Hazards, "The GEOHealth program is an innovative and productive collaboration that leverages investment of funding partners and the needs of academic, research and practice agencies around the world to respond to these difficult issues." E-waste has become a recent priority topic of the GEOHealth program, and Ghana has played a central role in these emerging hub research projects. In general, the West Africa GEOHealth project has helped provide the research support needed to extract and organize bioinformation on e-waste workers.

E-waste biometrics beyond bodily fluids

What I am calling *bioinformatic extraction* here involves the ways in which labor practices, as forms of bioeconomic life, are used as data points and biometric measures to better understand e-waste worker exposures to toxic substances. One research focus in emerging e-waste epidemiology is turning towards interventions where workers are actually wearing and embodying science, wearing observational and measurement tools to build e-waste toxics knowledge. Efforts to "track" e-waste worker behavior in the informal e-waste sector in Ghana has been the focus of recent environmental health science. Recently, researchers from the University of Michigan "designed and applied a method for objectively deriving time-activity patterns from wearable camera data and matched images with continuous measurements of personal inhalation exposure to size-specific particulate matter (PM) among workers [at Agbogbloshie]" (Laskaris 2019: 829).

Tracking the time-activity patterns of workers, these scientists contend, is not only a method that compliments air pollution sampling but is considered necessary for actually knowing the exposure situations confronted in a particular work environment. In this way, this recent study is stretching the range and content of bioinformatic collection practices and data points. Time-activity data is now becoming a biometric beyond blood, a derivative bioeconomic measure to advance e-waste epidemiological knowledge of the risks Ghana's e-waste workers face every work day.

This team of researchers created a "visual activity dictionary," a code-book for time-lapse images captured by cameras worn by e-waste workers. From these cameras, the researchers collected over 35,000 images over 170 work shifts. The images capture what the researchers call "time-activity patterns," or what workers actually do during the work day. The images collected were classified (i.e., sitting, walking, burning cables, smoking, eating, etc.) and synthesized with air pollutant exposure data, or PM2.5 (small particulate matter). To collect the PM2.5 concentrations, the researchers had the e-waste workers wear a backpack with a minute-by-minute PM detection device to measure real-time work activity exposures to toxic substances.

Going beyond the extraction of blood, semen, and maternal urine to measure lead and cadmium "body burdens"[6] is a new direction for e-waste epidemiology. What these new biometrics of e-waste labor risk add, researchers argue, is a less intrusive and burdensome method of data extraction:

> Wearable camera data ... eliminate the participant burden and literacy requirements associated with workers keeping active time-activity diaries with 5 or 15-min resolutions; such high-resolution diaries may be required in job settings with frequent task changes and acute exposures.
>
> (Laskaris et al. 2019: 832)

But what exactly do these new biometrics and "time-activity data" tell us about e-waste environmental health risk as it is socially and economically experienced? The answer is: very little. This is perhaps why the researchers note that future research methods "should involve an iterative process between workers, local leaders, and multidisciplinary teams, including engineers, exposure experts, epidemiologists, and *social scientists*" (Laskaris et al. 2019: 840, emphasis added). The assumption of e-waste epidemiologists in Ghana is that this population is made up of bodies navigating chemical exposures, but if social science engagement is actually taken seriously, as some have suggested (Little 2019, Little and Akese 2019, Akese and Little 2018), the biological frameworks guiding e-waste epidemiologists working in Ghana will surely need to be upgraded and informed by more critical social science engagements with "the body." So far, there has been no direct biopolitical critique of e-waste in Ghana and China, even despite the fact that all e-waste studies to date deal with or involve a focus on working *populations*, laborers situated in states with biosocial power, and bodies (as biodata) being studied and governed by epidemiological sciences and global health research teams and science partnerships.

What matters most is that the ongoing bioinformatic extractions in e-waste recycling sites is occurring alongside a social reality that complicates how situated biologies and laborers are themselves understood. One factoid of the local situation is that e-waste workers, whether in China or Ghana, are attempting to sustain livelihoods. Just as e-waste health studies were beginning to emerge in Ghana, it is necessary to understand the fact that e-waste scavenging is a wage-based livelihood strategy (Oteng-Ababio 2012). These livelihoods are linked to certain modes of extraction. As Grant would contend,

> We need a new and different perspective to capture these phenomena and the growing urbanization of mining that extends the modes of extraction via the recycling of e-waste, concentrated in particular city locales, and tied to international scrap circuits.
>
> (Grant 2016: 21)

Additionally, and to further showcase the ways in which e-waste epidemiology is a matter of bioeconomics, the epidemiological focus on toxic bioaccumulation necessarily integrates everyday labor data. Capturing this labor-bioinformation can hardly be a snapshot, since engagement in e-waste labor is more long-term than previously thought. Both in China and Ghana, the biolabor situation in these e-waste hot-zones requires instead a focus on life course exposures to capture epigenetic signatures of e-waste bioeconomies. This situation, which is supported by recent research findings, casts a new light on the importance of longitudinal occupational and environmental exposure studies guided by the collection of biometrics, especially the bioaccumulation of contaminants over extended periods of time.

Conclusion

Following the lead of medical anthropologist Margaret Lock, efforts to re-theorize the complexities of e-waste bioinformatic extraction must call attention to "situated" occupational and environmental exposures, to situated toxicity. E-waste workers, from Guiyu to Agbogbloshie, continue to offer a new materiality (and new biodata) to e-waste science studies, and a focus on "local biologies" and "situated biologies" provides an important addition to current and future anthropological perspectives on and critiques of global e-waste epidemiology. According to Lock, "local biologies" is a notion that "refers to the manner in which biological and social processes are permanently entangled throughout life" (Lock 2017: 8). This "life" in the world of e-waste workers, from Ghana to China to India, is one also suspended by what I have called elsewhere "technoecobiopolitics" (Little 2012)—a biopolitics entangled by techno-environmental disruption and contamination, and source of environmental health governance and science. If relations of laboring bodies and e-wastes are taken seriously, and if an anthropology of e-waste bioinformation is to become a springboard for an emergent eco-biopolitical theory, tech-industry power and politics of electronic discard governance become integral to any sustained critique. This technoecobiopolitics involves recognition of global political ecologies of e-consumption. For example, according to the U.S. Bureau of Economic Analysis, in the U.S. alone, consumers spent $71 billion on telephone and communication equipment in 2017, nearly five times what they spent in 2010 even when adjusted for inflation (Semuels 2019). There are countless examples of swelling e-consumption globally, but e-waste labor makes up the underlying toxic biopolitics of this corporate global electronics industry.

E-waste biopolitics reconfigures the bioeconomic relations of toxic substances and e-waste labor. This biopolitics is also aware of its own flexibility and situatedness, and therefore intentionally redirects attention to global bioeconomics, to related populations (related workforces) in e-waste hot-zones, like those in China and Ghana. Attending to this synthesis and redirection might help further expose the place and value of more "protective biopolitics of poison" (Tousignant 2018: 10) and how problems of e-waste risk are a consequence of a broader, more global political economy and ecology of toxic labor. Some epidemiologists are hinting at this more "protective" biopolitics of e-waste labor, even if they are not exactly calling it that or frame their practice in this new, twisted way. For example, a recent e-waste health science study involving partners from the U.S., Canada, and Ghana, have begun documenting the frequencies of e-waste workers' use of personal protective equipment while working with e-waste (Burns et al. 2019). What collecting evidence of bio-protection among workers will actually lead to is unclear, but it surely highlights the use of bioinformation to measure some degree of self-protection.

In this chapter, I have noted that to bring labor into global e-waste epidemiology demands a more direct interest in the social and bioeconomics of e-waste labor. This attention to bioeconomics has clear ties to Foucault, who wrote that

> to bring labor into the field of economic analysis, we must put ourselves in the position of the person who works; we will have to study work as economic conduct practiced, implemented, rationalized, and calculated by the person who works. What does working mean for the person who works? *What system of choice and rationality does the activity of work conform to?*
> (Foucault 2008: 223, emphasis added)

If bioinformatics figure in our critical understandings of toxic e-waste labor in Ghana and China, or anywhere e-waste recycling occurs, we must reckon with the complexities of economic and environmental health conditions, and how these conditions directly inform how e-waste recycling labor is practiced. In both Ghana and China, the toxic e-waste situation has led to an emergent system of technoscience and expertise involving organized bioinformatic extraction to build and stabilize e-waste epidemiological science. Surely other global health and epidemiological sciences engage in similar forms of biodata extraction, but the turn to e-waste recycling labor is recent. In my view, seriously engaging in a bioinformatic anthropology of this emergent global health concern calls for a dual focus on new political ecologies of e-waste (Little and Akese 2019) and the "geographies of science" (Livingstone 2003) activated and positioned to stabilize and scientize global e-waste health epistemics. As it turns out, knowing about the risks of e-waste labor in China and Ghana, involves knowing more than just toxic substances, dose-response relations, and certainly knowing more than simply "waste." It involves a much more complex situation of cultural relations and materialities, a process of learning how these relations and materialities are in fact situated for and by scientific practice itself.

As this chapter has tried to show, e-waste workers are, for sure, situated biologies, but they have become simultaneously objects (and research participants) of environmental and occupational health science. These workers, in other words, have become agents of toxic labor and offer data points of bodily extraction and surveillance. Additionally, the focus on worker "cohorts" and cohort exposures in e-waste health studies calls for a rethinking of the complex ways in which these working populations make up a biopolitical environment where workers often experience different relations to science in general and global health research projects in particular. This is especially important when considering real differences in socio-political context or when confronting variegated anatomies of power (Butchart 1998). For example, e-waste health science in Ghana—a presidential representative democratic republic—is largely driven by academic institutions and NGOs, while in contemporary authoritarian China, science is under strict "party-state controls" (Greenhalgh and Zang 2020: 9), even if academic scientists are involved in designing and carrying out this research. In this way, efforts to develop a "global" health science of e-waste involves similar and divergent approaches to understandings of science itself. In light of the critique that globalization is complex, disjointed, and marked with friction (Tsing 2005), our understanding of how "global" e-waste health science is made and validated matters, especially for a critical anthropology of e-waste science that explicitly reappraises the place of critique itself (Fassin and Harcourt 2019). What the cases in Ghana and China highlight is that the local context and situated biologies in focus, the worker cohorts and bioeconomic situation in focus, play a significant role in shaping the framing of research questions, what experimental methods are used, the style of investigation, the process of recruitment, and how toxics' risk and uncertainty are realized and theorized. Each research project, each bioinformatic hub, attempts to build e-waste health evidence based on site or setting-specific knowledge and authority (Knorr-Cetina 1999, Pickering 1992) where certain biologies and biodata—blood, semen, and maternal urine—become epidemiologically significant. In this way, the very epistemics of e-waste toxicity becomes a matter of actual "emplacement" (Reno 2011) whereby the production of a global e-waste epidemiological science simply reinforces and further necessitates the value and need for this site-specific bioinformation.

No doubt, the current theoretical landscape is wide open. Some recent e-waste scholarship is starting to attend to questions of bio-citizenship, with a particular focus on how e-waste disputes link to the embodiment of global disparities caused by neoliberalism and austerity (Sparke 2017). But, if what is on the horizon for e-waste epidemiology is a global health science practice informed by toxicogenomics and epigenetics,[7] the effort likely involves the scaling up of bioinformatic extractions and development of more science hubs. As this trend develops, anthropologies of bioinformation can guide these global health science efforts by encouraging a renewed focus on biosocial materiality and toxic vitalism to more fully comprehend the actual global health equity[8] dynamics of e-waste. As it currently stands, e-waste epidemiology is firmly anchored to and informed by certain biological reductionisms of environmental and occupational health science, even if, as noted here, attention to e-waste labor (or economic) behavior is integrated in some recent studies. For sure, health study "hubs" are vibrant economic zones, and occupational health studies focusing on labor "activity" are implicitly recognizing how environmental epidemiology is more than toxic exposures. These studies together fold in an interest in labor "activity" that highlights the experience of workers navigating their everyday neoliberal economic lives, lives poisoned by "vital infrastructures of labor" (Fredericks 2018). In a sense, it doesn't seem that radical to rethink e-waste epidemiology as a field that intentionally targets (even desires) toxic bodies, as this corporeal contamination and biodata provide the actual infrastructure of evidence needed to make environmental epidemiological claims about e-wastes' toxicity.

Despite this recent turn to labor experience more directly, the fact is, bioscience momentum is on the side of epidemiological experts, like GEOHealth scientists. Both in Ghana and China, little interdisciplinary science on e-waste health is being done, which further exemplifies the continued fortification of specific disciplines engaged in the production of e-waste health epistemics. This is likely due to the general toxics fixation of environmental epidemiology, a field dominated by a focus on exposures and health outcomes or risks of those toxic exposures. An enduring challenge for the bioinformation turn in anthropology is to confront, track, and critique these different emerging bioscience practices, especially as they get reassembled and globalize. Whether in China or Ghana, we must commit to creative synthesis and critique to actually come to terms with emerging e-waste bioinformatics. This may call for a synthesis of e-waste anthropology and e-waste epidemiological science in a way that builds an actual biopolitics of e-waste attuned to global transformations in e-waste economies, a research effort that will ultimately need to include

> the conversation of global resources (saving the environment, reincorporating materials already enmeshed in the global material system, and turning residuals into resources) and providing local [and healthy] livelihood opportunities (although in its present form, scavenging and informal processing are far from decent work).
>
> (Grant 2016: 27)

Avoiding this scientific synthesis only flattens the potential for a critical biopolitics of e-waste informed by new and emerging anthropologies of bioinformation.

Notes

1 I have been conducting ethnographic fieldwork on e-waste in Ghana since 2015, so my knowledge of the Ghana e-waste situation is far richer than what I have learned of e-waste studies in China from a distance.

2 The first project to really track e-waste trades was in 2002, with the Basel Action Network's documentary film project *Exporting Harm: The High-Tech Trashing of Asia*, a short documentary film about "dumping" electronics overseas, with a special focus on e-waste in Guiyu, China.

3 Based on my own fieldwork on e-waste labor in Ghana (Little forthcoming, Little 2019, Little and Akese 2019, Akese and Little 2018), women continue to be sidelined in public health research, as e-waste and the scrap metal economy there is a strongly male-dominated industry and market.

4 As noted in Concha-Barrientos et al. (2004), in 2002, the World Health Organization (WHO) completed an assessment of the risk factors contributing to the global disease burden of occupational noise.

5 GEOHealth funding partners, in addition to the Fogarty Center, include the U.S. National Institutes of Health, National Cancer Institute, National Institute of Environmental Health Sciences, the National Institute for Occupational Safety and Health within the Center for Disease Control and Prevention, Canada's International Development Research Centre, and the Clean Cooking Alliance, which is providing supplemental funding for research and training focused on household air pollution.

6 This term is almost universal in environmental epidemiology discourse, and refers to the burden caused by bodily accumulation and absorption of chemical substances.

7 For anthropological insights on toxicogenomics and epigenetics, see Fortun and Fortun (2005) and Lamoreaux (2016).

8 In a recent lecture at Duke University's Sanford School of Public Policy, Paul Farmer noted that "all public health now is or should be about global health equity," and the same could be said for global health projects focused on e-waste.

References

Adams, Vincanne, ed. 2016. *Metrics: What Counts in Global Health*. Durham, NC: Duke University Press.

Akese, Grace Abena and Peter C. Little. 2018. Electronic waste and the environmental justice challenge in Agbogbloshie. *Environmental Justice* 11: 77–83.

Aich, Nirupam*et al.*2020. The hidden risks of e-waste: perspectives from environmental engineering, epidemiology, environmental health, and human–computer interaction. In *Tansforming Global Health*. K. Smith and P. Ram, eds. Pp. 161–178. Amsterdam: Springer.

Armah, Frederick Ato*et al.*2019. Assessment of self-reported adverse health outcomes of electronic waste workers exposed to xenobiotics in Ghana. *Environmental Justice* 12 (2): 69–84.

Burns, Katrina N., Stephanie K. Sayler, and Richard L. Neitzel. 2019. Stress, health, noise exposures, and injuries among electronic waste recycling workers in Ghana. *Journal of Occupational Medicine and Toxicology* 14 (1):1–11.

Burns, Katrina N.*et al*.2016. Heart rate, stress, and occupational noise exposure among electronic waste recycling workers. *International Journal of Environmental Research and Public Health* 13: E140.

Butchart, Alexander. 1998. *The Anatomy of Power: European Constructions of the African Body*. London and New York: Zed Books.

Caravanos, J., E. Clark, R. Fuller*et al*.2011. Assessing worker and environmental chemical exposure risks at an e-waste recycling and disposal site in Accra, Ghana. *Journal of Health and Pollution* 1: 16–25.

Chan, Janet K. Y.*et al*.2007. Body loadings and health risk assessment of polychlorinated dibenzo-p-dioxins and dibenzofurans at an intensive electronic waste recycling site in China. *Environmental Science and Technology* 41: 7668–7674.

Chen, Aimin, Kim N. Dietrich, Xia Huo, and Shuk-mei Ho. 2011. Developmental neurotoxicants in e-waste: an emerging health concern. *Environmental Health Perspectives* 19 (4): 431–438.

Concha-Barrientos, M., D. Campbell-Lendrum, and K. Steenland. 2004. *Occupational Noise: Assessing the Burden of Disease from Work-Related Hearing Impairment at National and Local Levels*. Geneva: World Health Organization. WHO Environmental Burden of Disease Series, No 9.

Davis, J-M., G. Akese and Y. Garb. 2018. Beyond the pollution haven hypothesis: where and why do e-waste hubs emerge and what does this mean for policies and interventions? *Geoforum* 98: 36–45.

Fassin, Didier and Bernard E.Harcourt, eds. 2019. *A Time for Critique*. New York: Columbia University Press.

Feldt, T., J. N. Fobil, J. Wittsiepe, M. Wilhelm, and H. Till. 2014. High levels of PAH-metabolites in urine of e-waste recycling workers from Agbogbloshie, Ghana. *Science of the Total Environment* 467: 369–376.

Fortun, Kim and Mike Fortun. 2005. Scientific imaginaries and ethical plateaus in contemporary U.S. toxicology. *American Anthropologist* 107 (1): 43–54.

Foucault, Michel. 2008 [1979]. *The Birth of Biopolitics: Lectures at the Collége de France*. Translated by Graham Burchell. New York: Picador.

Fredericks, Rosalind. 2018. *Garbage Citizenship: Vital Infrastructures of Labor*. Durham, NC and London: Duke University Press.

Ghana Health Service. 2017. *2016 Annual Report*. June.

Grant, R. and M. Oteng-Ababio. 2012. Mapping the invisible and real 'African' economy: urban E-waste circuitry. *Urban Geography* 33: 1–21.

Grant, Richard. 2016. The "urban mine" in Accra, Ghana. In *Out of Sight, Out of Mind: The Politics and Culture of Waste*, Christof Mauch, ed. *RCC Perspectives: Transformations in Environment and Society* 1: 21–29.

Greenhalgh, Susan and Li Zhang. 2020. *Can Science and Technology Save China?*Ithaca, NY: Cornell University Press.

Huo, Xia, L. Peng, X. Xu, L. Zheng, B. Qiu, Z. Qi, B. Zhang, D. Han, and Z. Piao. 2007. Elevated blood lead levels of children in Guiyu, an electronic waste recycling town in China. *Environmental Health Perspectives* 115 (7): 1113–1117.

Isenhour, Cindy and Joshua Reno. 2019. On materiality and meaning: ethnographic engagements with reuse, repair, and care. *Worldwide Waste: Journal of Interdisciplinary Studies* 2 (1): 1–8.

Kaza, S., L. Yao, P. Bhada-Tata and F. Van Woerden. 2018. *What a Waste 2.0: A Global Snapshot of Solid Waste Management to 2050*. Washington, DC: World Bank.

Kim, Stephanie*et al.*2019. Metal concentrations in pregnant women and neonates from informal electronic waste recycling. *Journal of Exposure Science and Environmental Epidemiology* 29: 406–415.

Kim, S. S. 2020. Birth outcomes associated with maternal exposure to metals from informal electronic waste recycling in Guiyu, China. *Environment International* 137: 105580.

Knorr-Cetina, K. 1999. *Epistemic Cultures: How the Sciences Make Knowledge.* Cambridge, MA: Harvard University Press.

Meloni, Maurizio, Simon Williams, and Paul Martin. 2016. *Biosocial Matters: Rethinking Sociology-Biology Relations in the Twenty-First Century.* West Sussex: Wiley Blackwell.

LaDou, Joseph. 2006. Occupational health in the semiconductor industry. In *Challenging the Chip: Labor Rights and Environmental Justice in the Global Electronics Industry,* T. Smith, D. A. Sonnenfeld, and D. N. Pellow, eds. Pp. 31–42. Philadelphia, PA: Temple University Press.

Laskaris, Zoey*et al.*2019. Derivation of time-activity data using wearable cameras and measures of personal inhalation exposure among workers at an informal electronic-waste recovery site in Ghana. *Annals of Work Exposures and Health* 63 (8): 829–841.

Lamoreaux, Janelle. 2016. What if the environment is a person? Lineages of epigenetics in a toxic China. *Cultural Anthropology* 31 (2): 188–214.

Leong, Apo and Sanjiv Pandita. 2006. "Made in China": electronics workers in the world's fastest growing economy. In *Challenging the Chip: Labor Rights and Environmental Justice in the Global Electronics Industry,* T. Smith, D. A. Sonnenfeld, and D. N. Pellow, eds. Pp. 55–69. Philadelphia, PA: Temple University Press.

Lepawsky, Josh. 2018. *Reassembling Rubbish: Worlding Electronic Waste.* Cambridge, MA: MIT Press.

Little, Peter C. 2012. *Think Technoecobiopolitics: Reflecting on the Emerging Political Ecology of IBM's 'Smarter Planet' Mission.* Paper presented at the American Anthropological Association Annual Meeting, San Francisco, USA, November.

Little, Peter C. and Grace Abena Akese. 2019. Centering the Korle Lagoon: exploring blue political ecologies of e-waste in Ghana. *Journal of Political Ecology* 26 (1): 448–465.

Little, Peter C. 2019. Bodies, toxins, and e-waste labour interventions in Ghana: toward a toxic postcolonial corporality? *Revista de Antropología Iberoamericana* 14 (1): 51–71.

Little, Peter C. Forthcoming. *Burning Matters: Life, Labor, and E-Waste Pyropolitics in Ghana.* New York and London: Oxford University Press.

Livingstone, David N. 2003. *Putting Science in Its Place: Geographies of Scientific Knowledge.* Chicago and London: Chicago University Press.

Lock, Margaret. 2017. Recovering the body. *Annual Review of Anthropology* 46: 1–14.

Orisakwe, Orish Ebere*et al.*2019. Public health burdens of e-waste in Africa. *Journal of Health and Pollution* 9 (22): 1–12.

Oteng-Ababio, Martin. 2012. When necessity begets ingenuity: e-waste scavenging as a livelihood strategy in Accra, Ghana. *African Studies Quarterly* 13 (1): 1–21.

Pickering, Andrew, ed. 1992. *Science as Practice and Culture.* Chicago: University of Chicago Press.

Renfrew, Daniel. 2018. *Life Without Lead: Contamination, Crisis, and Hope in Uruguay.* Berkeley: University of California Press.

Reno, Joshua. 2015. *Waste Away: Working and Living with a North American Landfill.* Berkeley: University of California Press.

Reno, Joshua. 2015. Waste and waste management. *Annual Review of Anthropology* 44: 557–572.

Reno, Joshua. 2011. Beyond risk: emplacement and the production of environmental evidence. *American Ethnologist* 38 (3): 516–530.

Schulz, Yvan. 2015. Towards a new waste regime? Critical reflections on China's shifting market for high-tech discards. *China Perspectives* 3: 43–50.

Schulz, Yvan and Anna Lora-Wainwright. 2019. In the name of circularity: environmental improvement and business slowdown in a Chinese recycling hub. *Worldwide Waste: Journal of Interdisciplinary Studies* 2 (1): 9. doi:10.5334/wwwj.28.

Semuels, Alana. 2019. The world has an e-waste problem. *Time*, May 23.

Singer, Merrill (Ed.). 2016. *A Companion to the Anthropology of Environmental Health.* West Sussex, UK: Wiley Blackwell.

Srigboh, R.K.*et al.*2016. Multiple elemental exposures amongst workers at the Agbogbloshie electronic waste (e-waste) site in Ghana. *Chemosphere* 164: 68–74.

Standaert, M. 2015. *China's Notorious E-waste Village Disappears Almost Overnight.* Seattle, WA: Basal Action Network.

Tichenor, Marlee. 2016. The power of data: global malaria governance and the Senegalese data retention strike. *Metrics: What Counts in Global Health.* Vincanne Adams, ed. Pp. 105–124. Durham, NC: Duke University Press.

Tousignant, Noémi. 2018. *Edges of Exposure: Toxicology and the Problem of Capacity in Postcolonial Senegal.* Durham, NC: Duke University Press.

Tsing, Anna Lowenhaupt*et al.*2017. *Arts of Living on a Damaged Planet: Ghosts and Monsters of the Anthropocene.* Minneapolis: University of Minnesota Press.

Tsing, Anna Lowenhaupt. 2005. *Friction: An Ethnography of Global Connection.* Princeton: Princeton University Press.

Schulz, Y. and Lora-Wainwright, A. 2019. In the name of circularity: environmental improvement and business slowdown in a Chinese recycling hub. *Worldwide Waste: Journal of Interdisciplinary Studies* 2 (1).

Sparke, Matthew. 2017. Austerity and the embodiment of neoliberalism as ill-health: towards a theory of biological sub-citizenship. *Social Science and Medicine* 187: 287–295.

Tang, C.*et al.*2010. Heavy metal and persistent organic compound contamination in soil from Wenling: an emerging e-waste recycling city in Taizhou area, China. *Journal of Hazardous Materials* 173: 653–660.

Walkover, Lily. 2016. When good works count. In *Metrics: What Counts in Global Health.* Vincanne Adams, ed. Pp. 163–180. Durham, NC: Duke University Press.

Wang, Yan*et al.*2018. Genomic instability in adult men involved in processing electronic waste in northern China. *Environment International* 117: 69–81.

Wong, M. H.*et al.*2007. Export of toxic chemicals: a review of the case of uncontrolled electronic-waste recycling. *Environmental Pollution* 149: 131–140.

Wong, C. S.*et al.*2007a. Evidence of excessive releases of metals from primitive e-waste processing in Guiyu, China. *Environmental Pollution* 148: 62–72.

Wong, C. S.*et al.*2007b. Trace metal contamination of sediments in an e-waste processing village in China. *Environmental Pollution* 145: 434–442.

Zeiderman, Austin. 2015. Spaces of Uncertainty: Governing Urban Environmental Hazards. In *Modes of Uncertainty: Anthropological Cases*, Limor Samimian-Darash and Paul Rabinow, eds. Pp 182–200. Chicago: University of Chicago Press.

Zheng, Liangkai, K. Wu, Y. Li, Z. Qi, D. Han, B. Zhang, C. Gu, G. Chen, J. Liu, S. Chen, X. Xu, and X. Huo. 2008. Blood lead and cadmium levels and relevant factors among children from an e-waste recycling town in China. *Environmental Research* 108 (1): 15–20.

9 Seeing like an airport

Towards interoperability in contemporary security

Mark Maguire and Eileen Murphy

Certain forms of knowledge and control require a narrowing of vision. The great advantage of such tunnel vision is that it brings into sharp focus certain limited aspects of an otherwise far more complex and unwieldy reality.

James Scott, *Seeing like a State* (1988: 11)

Introduction

In the three decades since the publication of James Scott's influential *Seeing like a State* we have been encouraged to "see like" an autocrat, an international organization, an oil company, a market, a border, and, more than once, a city (Jones 2015, Broome and Seabrooke 2012, Ferguson 2005, Fourcade and Healy 2017, Rumford 2012, Magnusson 2011, Amin and Thrift 2016). The list is long, but, surprisingly, it does not include the modern airport. This is striking because Scott's aim was to show how modern state institutions are animated by codification and quantification, rendering persons and things legible in standardized space. What better example than an international airport? During the 1940s, the United Nations International Civil Aviation Organization (ICAO) began the process of turning the world's air into a navigable aerial world, and from that point onwards airport terminals became modernist triumphs of visibility and legibility. Why, then, have we not been encouraged to see like an airport?

While aviation modernism has long been celebrated in bookshelves and on coffee tables, the relevant critical social-scientific literature remains small and rather dismal. Indeed, anthropologists and others tend to describe airports as the environmentally destructive strongholds of capitalism, as gated spaces for the kinetic elite, or as the embodiment of Orwellian surveillance. However, few studies are based on meaningful ethnographic access. Indeed, most of the well-cited texts openly admit to non-trivial methodological limitations (e.g. Augé 1995, Chalfin 2008, Adey 2010). Interestingly, in *Politics at the Airport* (2008), Mark Salter celebrates this inadequacy:

> The airport is a messy system of systems, embedded within numerous networks and social spheres. ... Airports are ... a network of networks that

DOI: 10.4324/9780367810030-9

include social, economic, and political actors with differing preferences, goals, logics, intentions, and capabilities. Rather than seek the single key idea of the airport, critics should instead focus on the resultant system— essentially, systems of rule without systematizers, convergence without coordination.

(Salter 2008: xiv)[1]

But does all that is solid really just melt into the air of innumerable assemblages? What if we looked to the actual trends in visualization within the contemporary airport, most especially the ongoing drive towards interoperability?

To illustrate, some years ago we attended the annual Security Research, Innovation and Education Event (SRIEE) in Tallinn, Estonia. This event opened with energetic discussion of "dematerialized borders". Then Estonian Prime Minister Jüri Rata imagined a near-future in which people experienced borders through "100% … electronic solutions". Striking a similar tone, Edgar Beugels, R&D Director of FRONTEX, imagined a near-future in which someone could depart their home in Singapore and arrive in Europe having travelled through a seamless "corridor" composed of interoperable sensors and systems. Tempting though it is, these statements cannot be dismissed as the Bentham-like dreams of politicians and bureaucrats on junkets. On the contrary, the near-future airport is already available in architectural form. Dubai International Airport, for example, is experimenting with a smart corridor that captures facial and other identifiable characteristics while appearing to be a virtual, walk-through aquarium, no less. Here we have the near-future airport as relations coded for speed and efficiency and enabled by seamless, interoperable (and potentially entertaining) security platforms. These platforms can "see", like an airport as it happens, albeit in the form of multiple glances more so than a steady gaze.[2]

Today, seeing like an airport is not an academic formulation to explain modernity but, rather, a specific problematization that is attracting attention, expertise and solutions. This problem is normally cast as fragmentation and increasingly dealt with under the rubric of interoperability. We explore this problematization by first reviewing the main theories of visualization available in the social sciences, showing the specific ways in which they illuminate some aspects of contemporary airports but not others. Thereafter, we discuss two spheres of the contemporary international airport experience, namely counter-terrorism policing, and the emergent data-border infrastructure. We will show the remarkable efforts in these soon-to-be-interconnected spheres to develop a near-future of seamless interoperability. We conclude by discussing some of the important implications of the transformations that are taking place, fore-grounding the role of para-ethnographic knowledge.

Our contribution here draws from sustained research on airports, counter-terrorism and border security in several European contexts. Murphy's experience stretches from ethnographic research on the medical-security nexus at borders to the societal impacts of automated border control (see Murphy and Maguire 2015). Her current work explores the role of eu-LISA, the European

Agency for the operational management of large-scale IT systems in the area of freedom, security and justice, especially questions of infrastructure, interoperability, and the politics of data management. Maguire's research stretches from ethnographies of future security technologies to counterterrorism operations (Maguire 2014, Maguire and Fussey 2016, Maguire 2018). His contribution here draws from two six-month (late 2011 to early 2012 and late 2015 to early 2016) ethnographic projects with two international airports in the British Isles, both of which involved assessing the effectiveness of international anti-terrorism models in the light of actual policing practices. Our collaborative work emerges from an unusually high degree of access, and such access depends on a willingness to work para-ethnographically with experts as counterparts with shared professional recognition (see Holmes and Marcus 2005). We are committed to effective critique, which implies a commitment to ethnography as a space wherein rigorous assessment of the state of things and the state of the world is both possible and necessary.

Shadow of the tower

How exactly do social scientists conceptualize the visual elements of airports? Curiously, Jeremy Bentham's panopticon—the eighteenth-century diagram for a steel and glass prison structured around an inscrutable but all-seeing tower—remains the dominant image of ocular power. Scholars and activists tend to either adopt the concept wholesale, offer novel readings of it, or use it as a dramatic foil when setting out their own position. In all cases, Bentham's tower still casts a very long shadow indeed.

Some scholars studying airports argue that the panopticon is now a "reality" in the form of "biometrics, digital risk profiling ... closed circuit television and security checks" (see, *inter alia*, Soeters 2018: 81, Gordon 2004: 167). However, empirical studies of airports that set out to explore panoptic models find that the concept lacks any explanatory power (e.g. Coulton 2014). On a very practical level, airports are simply too large to be seen from any one centre—indeed, the question, as we describe it, is how to get multiple centres to converge. In 2018, for example, London Heathrow catered for 80 million passengers, and on its busiest day, at any one time, a quarter of a million passengers and tens of thousands of workers could be found on its 1,200 hectare campus. That same year, as part of an ongoing project on airport police training, Mark Maguire organized a "walk through" of a smaller airport in the British Isles which included international counterterrorism experts and senior European police. Even with such a powerful group on site, the CCTV control room was far from one's image of a centre of calculation and surveillance: new and old technologies were patched together, and one important CCTV monitor was obscured by a policeman's takeaway lunch.

But there are post-panoptic writers struggling to leave the shadow of the tower, many of whom focus on bioinformation. Mark Poster discusses the coming together of electronic surveillance and computer databases as the "superpanopticon" (see also Lyon 2008: 151, Bauman 2005). Didier Bigo offers a similarly inventive twist by proposing that the contemporary spatio-temporal disciplining of safe and normal spaces involves the risk-based exclusion of dangerous others, a "banopticon". Poster and Bigo's models are certainly provocative, but they work off a vision of the airport that is preoccupied with the movement of passengers across the border from landside to the secure or "clean" airside zone, a movement that requires one pass through a gauntlet of technologies and processes designed to sift, sort, grant access or exclude. But this does not resolve the problem of scale, nor does it take account of the vision of seamless interoperability. Thomas Mathiesen's "synopticon" emphasizes free-floating and media-based control (Mathiesen 1997, see also Hardt and Negri 2003: 196, 330). And, most famously, Haggerty and Ericson (2000) talk of "surveillant assemblages" to understand how bodies are rendered as code, separated into flows, and reassembled into "data doubles", which can then be examined and potentially targeted. Instead of the God's eye of Bentham's tower one has a multiplicity of vision machines. But what does this more polycentric image of the airport give us exactly? Recall here James Scott's (1988: 11) image of state "seeing" as the practical "narrowing of vision" that is needed to focus within a complex reality—or, for that matter, Bruno Latour's (2005: 181–187 *passim*) "oligoptica", the spaces where important microprocesses are staged and can been studied. But, again, the para-ethnographic is important here. In defiance of approaches that imagine panoptic towers, or specific rooms where power is centred, or even those who see a confusing multiplicity of assemblages, the para-ethnographic does not take the world as it imagines it should be; rather, it takes the contemporary as an unfolding ratio of modernity that merits exploration as it unfolds. We begin by examining shifts in the policing of airports.

Seeing like counterterror police

On the morning of September 11, 2001, five hijackers boarded American Airlines Flight 11 on route to Los Angeles. By the time they took their seats, most of them had been flagged as a risk by CAAPS—the computer-assisted passenger screening system—and their leader had been tagged by various intelligence agencies as a "suspicious" person. Yet the five men moved seamlessly through crowds and past several layers of security before boarding Flight 11. Of course, systems can always take more investment, and processes can be made more robust. And so, in the years after American Airlines Flight 11 smashed into the World Trade Center North Tower, massive investments were made in first and second-generation biometric systems that linked identification documents to human bioinformation. Airports now bristle with new scanners, and algorithms search the near-future for suspicious patterns of behaviour. Yet, much still depends on so-called human factors. At the end of the day, even if assisted by technology, it is people who must narrow their vision to focus on threats. But how exactly does one see a threat in a public space heaving with many thousands of human beings?

Here is Charles Dickens's note on security vision in his short story "On Duty with Inspector Fields":

> Inspector Field's eye is the roving eye that searches every corner. ... Inspector Field's hand is the well-known hand that has collared half the people here. ... All watch him, all answer when addressed, all laugh at his jokes, all seek to propitiate him. This cellar company alone ... is strong enough to murder us all, and willing enough to do it; but, let Inspector Field have a mind to pick out one thief here, and take him; let him produce that ghostly truncheon from his pocket, and say, with his business-air, "My lad, I want you!" and all Rats' Castle shall be stricken with paralysis, and not a finger move against him, as he fits the handcuffs on!
>
> (Dickens 1851: 2)

Dickens exposes core problem of the modern: bureaucratic order—in this case policing—requires skill and experience to be harnessed and directed, but order is then tied to elements that are imponderable, that elude quantification. Moreover, through Dickens we also see that the enforcement of order relies upon an exchange between people that shows the contours of an agreement—it is not simply a matter of ocular power.[3]

After 9/11 it was obvious that techno-scientific systems and process alone could never secure an international airport, because the problem was not necessarily one of fake documents, poor database records or suspicious patterns of behaviour. Rather, the problem was how to bring systems and process into line with the visual regime of the police patrol—the challenge was and still is to link the Inspector's eye to the relevant IT system, to provide more-than-human security. In late 2001 and early 2002, experiments began in Boston Logan Airport and later in several European countries (earlier work took place in Israel) to develop and standardize a behavioural observation system (see Di Domenica 2011). At the same time, massive investment went into sensor-laden corridors and algorithms to detect physical signs of "malintent" (see Maguire and Fussey 2016), from sweating and eye movement to body temperature and heart rate. Yet, in case after case in which would-be terrorists were apprehended, the most effective defence seemed to be the skilled eye of today's Inspectors. But what does all of this look like in practice today?

In late 2015 and early 2016, Mark Maguire spent six months shadowing and interviewing airport police in a major airport in the British Isles. At the time, aviation security vibrated in response to the increasing threat to European airports posed by international terrorists. New recruits were being taken on, training programmes and technology adoption were expanding. The police were considering the adoption of body cameras, for example, but concerns over privacy regulations confined this and other technology solutions to "pilot projects". There were also major concerns about the deployment of heavily armed counterterror police. These were uncertain times for already-stretched police, and technology and training requirements often seemed like an additional burden to carry. Here is a senior policeman describing the regulatory and compliance milieu that his officers patrol in:

> We have to be compliant with all regulations and instructions and whatever from both Transport and from the Aviation Authority and from US Federal Aviation and all these organisations that control civil aviation. You are liaising with different organisations, I suppose, at a relatively senior level in terms of daily operations and stuff that might be happening. You know, you've got the police, you've got the Emergency Services both locally and nationally; we have our own [airport] fire service, you've got the city Fire Brigade, the health officials … I know it sounds ridiculous, but we would have had stuff coming in on aircraft—there was a big scare about a virus, and all these type of things—so I suppose the airport environment is small and people think nothing happens here, and whatever, but there is a very, very international … So you do have to have good links and good communication with all these other agencies. And it's all happening, all the time, coming at you.
>
> (Interview recorded 2016)

In defiance of images of paramilitary style policing in panoptic airports, one instead finds evidence of a bureaucratic world of regulations, one that crashes into the sociological reality of tens of thousands of members of a diverse public. The para-ethnographic experience is dizzying, as to answer even a simple question an "informant" must reels off the names of dozens of organizations and bodies, each of whom has an unmemorable acronym, to explain a regulation or a process. And yet, this complex, fluid milieu is experienced as a human realm of personal contacts, interactions, and drama. After a long shift with a two-person patrol, Mark discussed the experience of policing a busy airport with two officers:

> You see it all, don't you? Everything. Last week a woman presented herself to me, and she had been sexually assaulted. She had been dropped off to the airport by the person who attacked her. That one for me now … was … hard. There's nothing that's going to train you, and no way to know what to say. I've been at suicides, and, again, nothing can prepare you, no real training except for your medical training, so you do everything you can to try to save the person. And then there's the violent people, like we saw today. I would say that age, and wisdom, and having now a family myself is the best training, because you learn to be smarter with people. It's not always best to go forward with people. Sometimes you take a step back and see if there's a better way to resolve things, to deal with something. *My greatest skill is not arresting people*, talking them down.
>
> (Interview recorded 2016 [our emphasis])

It is precisely into the woof and warp of everyday policing that new technologies are folded, each innovation promising legal and procedural compliance, transparency, and counterterror capability. Here is one seasoned officer describing the pilot introduction of body cameras:

> So you, as an officer, you're connected to the CCTV centre. [*They have the*] footage, and you've got your man, say, who you're looking at. Say you call in an assault. The officer goes, "Where was it?", "It was Level 2", And the control centre goes in, zones in on somebody, comes out, moves around and you can see him going. ... Then you'll have this [*camera*] on your vest now. You'll see and hear what happens when we approach the guy, and the CCTV will pull wide to capture everything.
>
> ...
>
> If you were going to a situation that's potentially violent and the person is going to ramp it up, if they think that they are being recorded they are much less likely if they are in the correct frame of mind, and not on drugs or not alcohol-induced, they are much more likely to respond to, "Listen, you're on camera, everything you say is going to be recorded". They are going to behave themselves a lot easier than if they weren't on camera. So this is for us a conflict resolution tool and would make the job easier, it would make the officer's life easier and would support any use of law through force that the officer found that they had to engage in as well, because this is the background to it, this is exactly what happened, it's not what I wrote down about 10 minutes later, it's live footage, there you go! So it supports the officer's behaviour and encourages as well correct behaviour in an officer because the officer himself knows that he's on camera too
>
> (Interview recorded 2016).

It is not surprising, then, that the next wave of "bleeding-edge" technology involves wearable augmented reality (AR) kit that brings smart environments and devices together with human and non-human interfaces. For some, this presents a frightening vista (see Hill 2020), and yet everything seems so sensible in the advanced spaces of liberal democracies. Why wouldn't one seek sensible connectivity between technology and policing? Much has already been said about mission and function creep, and resulting path dependencies, in the world of security, but not enough has been said about the extraordinary pressure driving towards interoperability. Why is this?

So far, we have discussed transformations in the airport policing realm with a particular emphasis on human-technology connectivity and the visual. However, airport policing is not a well-researched topic, and it seems to bring together two distinct realms—the airport as modern temple to mobility, and security as the modern obsession with control, especially control over mobility (the unwanted kind). The airport presents itself as an exceptional space, and indeed it is legally exceptional. But, though categorized as critical infrastructure, airports more often seem to be desperately mundane places to work—excitement quickly fades to the grey of bureaucratic modernity. And, though the police have exceptional powers to stop, search and arrest, and even to deploy deadly force, much of their quotidian experience is characterized by non-exceptional routine. The coordination of airport policing and, consequently, the absorption of new technology is more a matter of human capital than panoptic investment.

Police spend many of their shifts in the woof and warp of airport terminals, but as one walks with a foot patrol one leaves the glass and steel buildings and crosses into parking structures, over to office blocks, along warehouse space, even, perhaps, a small park. There are dark spaces, hidden stairwells and doors that are known only to particular service workers. There are massive areas with no camera coverage at all. And, when one joins a mobile patrol in a car or SUV the sheer scale of an airport reveals itself. Data representations of airports show them to be, essentially, corridors that link different forms of transport to shopping areas and airplanes, but the police know other spaces where wildlife flourishes absent humans. Once again, technology floods into gaps. In the airport where Mark Maguire spent six months in 2016, the perimeter is now patrolled by drones that fly to and from a "nest" and are controlled by means of a smart watch on the wrist of a technology-laden police officer.

Gaps and shadows, dark or blank spots—the empire of security is not all that it seems. It is important, however, to remind ourselves that security, conceptually and practically, is about the toleration of uncertainty, unknowns, the acceptance of certain amounts of risk to appease the gods of efficiency and enterprise. In the realm of security, the question is not if gaps exist—they always do—but, rather, which gaps become problems in need of solutions, for whom, and why?

Inside the terminals there are dark spots too. Police forces are hierarchical bureaucracies, and individuals on one level may not know what happens above their "pay grade". Here is an undercover airport police officer describing the formative years of this training:

> I was the most hopeless guy for spotting shoplifters because ... well, we introduced plain clothes people then and you'd be dying to get into plain clothes, and I thought I'd be brilliant at it but, no, apparently, because I can be spotted, "That guy is definitely a copper over there!" no matter if I dressed like a down and out or a prince. ... I never had any success in plain clothes early on. So ... I started to bring in my own scruffiest clothes I could get, and I used to get a boarding card for myself, and I used to have all sorts of things, caps and glasses—Jesus, I was going around here like f**king Inspector Clouseau at times. But, again, very little success because I think first of all I suppose there was no training in that area so I was making it up as I go along myself and trying to formulate some sort of method of doing this.
>
> (Interview recorded 2016)

As time went on, he and his fellow officers began to rely more and more on intelligence reports that zeroed in on known criminals with suspicious travel patterns. CCTV cameras would pick them up wandering about the airport without having exited through customs. Of course, biometric documents and international databases can only do so much. Human skill and physicality remain important, as does "experience", but there is always the promise of more interoperability. In 2016, two police officers discussed the potential future as they saw it:

> What we do is down to experience; it's down to reading body language; a person's demeanour, and what they're carrying with them. There can be give away signs, like a little bag, two or three little bags is a sure thing. But then again the pickpockets are really clever. They know what we know— like us, they know what's normal in a particular area, and how to blend in.
>
> [One officer, looking out the window and down to the foyer points and says, "Look, he's a passenger, and *he* isn't!" He then calls in a report ...]
>
> I used to talk to everyone I could. We used to call that "profiling". A dirty word now. ... Now it's Control that is in my head, talking into my ear, saying there's something suspicious outside the McDonalds, or the camera is picking up something over near doorway 6. So now I have information before I get somewhere. So now your mind and your body are disconnected a bit. It's strange somehow. They want you to be listening, and now recording everything with a body camera, and it seems the more things your hang on your uniform and the more you look like *Iron Man*, well, the more you are in reality but at a slight distance.
>
> (Interview recorded 2016)

As we have written elsewhere, there are multiple security sites, from borders to policing, where ostensibly powerful (at least according to the social-scientific literature) human operators are being constituted as a problem to be remedied with socio-technical "solutions" (see Murphy and Maguire 2015). This is certainly true in the contemporary international airport, where police are festooned by more and more kit, and where the drive is to reengineer systems with the goal of interoperability. This trend is certainly visible in the realm of counterterror policing. Indeed, increasingly over the past several years, whatever the question is, the answer is greater interoperability. So, what exactly do we mean when we deploy this term in the context of counterterrorism?

Fundamentally, counterterrorism is about prevention and protection. Should a European airport suffer a terrorist attack, police will attempt to protect civilian life while alerting cooperating authorities as part of protocol. The alert, depending on severity and the region, is likely to reach counterterror forces, from elite firearms teams to black-clad Special Forces and emergency ordonnance units, who will assume responsibility under special powers legislation. Thus, there is vertical separation that requires ongoing effort. Then there is the pre-emptive work of intelligence and surveillance which generates horizontal separation of activities and visual regimens. In crude terms, from the perspective of counterterrorism the airport sees through a myriad prisms. Thus, as a police officer patrols through an airport, she passes disconnected lines of sight, separate technologies, and walks across different possible jurisdictions in emergency situations. Little wonder, then, that the multiplicity of systems and visual regimes is a problem that, so the experts hope, is to be resolved by interoperability. There is no point having an undercover unit if they cannot be called upon to spot terrorists, so the argument goes, or a surveillance system to prevent criminal damage if it cannot be pointed to a more dangerous threat—after all, whatever the question is, the answer is always greater interoperability.

Of course, one wonders if all of this talk about interoperability masks some other hidden process: perhaps it is a technical discourse hiding the emptiness of security theatre, or perhaps a technical mask behind which stands the commercial interests of information technology companies, or even the surveillance interests of powerful governments? Here we have proposed that the on-the-ground reality of airport security suggests that interoperability denotes a way of thinking, through which to "better" organize reality, and a mechanism through which those responsible for the airport can see others and, crucially, see themselves.

Interoperability, with Marie Kondo

But airports are, as we have already suggested, more than simply problems of and for mobility. Airports are shopping malls, office blocks, commercial hubs and, of course, they are borders. Just as we have traced a line along the human capital of the airport police, one can also look to the drive to interoperability among neighbouring forces. For the past two years, Eileen Murphy has worked on the problematization of large-scale information technology integration for European border control. Below, an IT expert narrates the waves of technical change through his own story:

> Yes, when I began my career back in the '70s as an immigration officer on the line at [names British airport], we didn't have any technology, it was a manual passport check, a check to see whether or not the person holding the passport looked like their passport photograph and a watch list check was conducted through a manual book by looking up someone's name on a list, a book, which was quite a clunky way on doing things. As we developed technology we moved towards automated checks against a watch list of suspect individuals. We then developed machine readable passports—you know passports that could be swiped and data could be checked electronically against watch lists. Then, increasingly lately, we've developed … biometrics where we're able to match people against their document using automated fingerprint recognition, or more likely nowadays facial recognition technology, and that saw the introduction of things like automated gates that people can use with electronic passports and also the development of targeting centers. As I said, I have a lot of experience with border security and, you know, it's important to us to make sure we had proper

data and analytics so that we could identify people who may be criminals or terrorists, suspect persons that were moving around the world and try to disrupt them. Yes, I have seen an explosion of technology really in the last 20 years or so, but it's increasing at such a rate now it's quite hard to keep up actually.

(Interview recorded 2018)

The text above helps to situate Murphy's interest in eu-LISA, and her para-ethnographic work with that Agency speaks to the unfolding efforts to "see" at the border.

Established in 2011, the European Union Agency for the operational management of large-scale IT systems in the area of Freedom, Security and Justice (eu-LISA)—hereafter referred to as the Agency—began operations in December 2012, taking over responsibility from the European Commission. The Agency is dispersed across a number of geographical sites, with its headquarters located in Tallinn, Estonia; its operational site located in Strasbourg, France; a business continuity site located in Sankt Johann im Pongau, Austria; and a Liaison Office in Brussels, Belgium. Currently, the Agency is responsible for the management of three large-scale IT systems in operation, SIS II, the VIS and Euro-dac. The Agency was established as a "long term solution for the operational management of large-scale IT systems, which are essential instruments in the implementation of the asylum, border management and migration policies of the EU" (eu-LISA website). In addition to managing the existing systems—and their ongoing legislative developments and recasts—the responsibilities of the Agency have expanded to include the development, implementation and operational management of three recently adopted systems, the Entry Exit System (ESS), the European Travel Information and Authorization System (ETIAS) and the European Criminal Record Information System for Third Country Nationals (ECRIS-TCN). Furthermore, the Agency has been tasked with developing interoperability between current and future systems. Formally adopted into regulation in 2019 (2019/818 and 2019/817) interoperability is defined as "the ability of information systems to exchange data and to enable the sharing of information" (COM (2016) 205 Final: 14). Responding to growing calls for border guards, airport police officers and relevant competent authorities to have access to the right information at the right time, interoperability is defined in relation to challenges within European databases. The introduction of a recent feasibility study on a future architecture of interoperability describes the situation thus:

In recent years, the areas of border management, internal security and migration management have been going through a major transformation, moving from the physical to the virtual world. Fragmentation among the existing EU information systems makes data access complex. This can lead to blind spots for law enforcement and other authorities, as it can be very difficult to recognize connections between data sets. Interoperability across the various information systems at EU level seeks to address this fragmentation. Through interoperability, systems will be able to supplement each other so that the appropriate authorities have access to the information they need, when they need it. Thus, interoperability between systems will help in tackling irregular migration, correctly identifying persons, fighting identity fraud and validating travel documents, and will ultimately contribute to a higher level of security in the area of freedom, security and justice in the EU.

(2019: 4)

Statements similar to this one are repeated across policy documents, office websites and reports. In a Communication from the Commission to the European Parliament and the Council, "Stronger and Smarter Information Systems for Borders and Security" (COM (2016) 205), the existing challenges of EU data management architecture are laid out, namely four main shortcomings: suboptimal functionalities of existing information systems; gaps in the EU's architecture of data management; a complex landscape of differently governed information systems; and, finally, a fragmented architecture of data management for border control and security (see 2016: 3). Technically, interoperability requires the coming together of multiple related elements: a Common Identity Repository, which is a shared holder of information about non-EU citizens, both biographic and biometric, stored in different systems; a Shared Biometric Matching Service, which is a tool for cross-checking biometric data and detecting links between information on the same person in different EU information systems; a Multiple Identity Detector, an automatic alert system capable of detecting multiple or fraudulent identities; and, finally, a European Search Portal, described as a "one stop shop" for carrying out simultaneous searches of multiple EU information systems. Interoperability is intended to not only remove information "silos" in relation to querying and searching the individual information systems, but it is also part of a drive to improve and standardize data input and use within the systems. All of this affects system designers and coordinators in multiple ways, but it also impacts on those authorized to carry out searches, such as police officers and border control agents. But it is the interoperable system that is foregrounded in terms of governance, not any individual agent.

At the 2019 annual eu-LISA conference titled, "The New Information Architecture as a Driver for Efficiency and Effectiveness in Internal Security", the keyword was interoperability. Richard Rinkins, Policy Analyst and Coordinator for Biometrics and Identity Management, DG Migration and Home Affairs opened by describing interoperability as similar to the central message in the successful Netflix TV series *Tidying Up with Marie Kondo*. He compared interoperability to decluttering and implementing a practice of storing like data with like data. Extending the reference, he went on to compare the efforts of looking for data within different systems to looking for socks, shoes, keys and glasses in your house. He even shared an amusing anecdote of his own troubles locating a textbook, which he eventually found in his refrigerator—his point being, if you can't easily find and therefore use data, it is of little use.

Of course, good practice and data "hygiene" are going to be considered as important, to be celebrated, valorized, but there is no naiveite in the discourse of these experts. There is, instead, a clear-sighted view that interoperability will assist in removing the shadowed areas and blind spots, but its implementation will be a significant task and one that will generate new problems. In illustrative terms, a former, very senior IT systems expert described his prominent role in European systems integration vis-a-viz the national operators who "own" the data:

> We have some knowledge on the national side but not to the extent that a national operator would have, because they're the ones running it, we are a little bit blind to that because it is their own thing. So, we know of things that affect us at the national side, but we certainly don't have a full picture by any means.
>
> (Interview 2018)

The key here is that like the policeman on patrol, systems security experts describe a world of fragmentation (once necessary because of specialization), with multiple visual regimes, that is in need of tidying, connecting and systematizing. Instead of convergence without coordination, here coordination is king. And as the police officer of the near-future walks her patrol route she will increasingly interact with connected systems that see, like an airport.

Conclusions

In the decades since the publication of James Scott's *Seeing like a State* much has changed in terms of how the critical social sciences approach bioinformation of all sorts, visual technologies, the relations between power, knowledge and seeing, and complex organizations. Yet, a blind spot remains when it comes to the airport, that most modern of modernist spaces. Anthropologists, it is plain, often fail to find meaningful access routes past security and up to the management suites of the contemporary—many exemplary studies openly admit as much (e.g. Augé 1995, Chalfin 2008). But lack of access is compounded by massive theoretical overstatement. Today, if we follow the social-scientific literature on airports we should expect to find Bentham-like ocular tyranny or loose visual assemblages; we should find powerful "superpanopticons" and targeted "data doubles" or "systems of rule without systematizers, convergence without coordination" (Salter 2008: xiv). There is an old saying that we know a strong wind is blowing when even the broken weathervanes point in the right direction. Here the theoretical weathervanes are not even detecting the strongest wind of change.

As the experience of police on the ground shows, technologies are swarming in airports, they are increasingly working together with human "capital".

Elsewhere in the airport, the lives of border guards and systems administrators are being actively challenged and changed under the star of interoperability. The work of eu-LISA shows the ways in which large-scale systems are being perceived, conceived and implemented. When police or border guards make an arrest, right now, they will be leaving a trace in an interoperable network, and as time goes on, that interdiction will itself become the subject of ever greater interoperability.

And what of the human subject in this near-future of seamless human to system interaction? What of the possibility of life beyond bioinformation code? From the outside looking in, either at large-scale IT systems or at policing—mindful that both invite a steady critical gaze—the temptation is to start with oversimplification. One is tempted to speak of surveillance or algorithmic rule as if that is external to the human experience, and thus something to be resisted. The para-ethnographic, because it suspends critique until criticism can be aimed accurately, opens a different set of questions about the airport and data-border today. The para-ethnographic is interested in the unfolding of the contemporary, and when it comes to the human and technical efforts to govern highly complex and data-saturated environments like the airport, we perceive a different starting point. The Rubicon was crossed long ago. In *The Human use of Human Beings* (1989 [1950]), Norbert Wiener, father of cybernetic system integration explains, with a warning,

> We have modified our environment so radically that we must now modify ourselves in order to exist in this new environment. We can no longer live in the old one. Progress imposes not only new possibilities for the future but new restrictions. It seems almost as if progress itself and our fight against the increase of entropy intrinsically must end in the downhill path from which we are trying to escape. ... The simple faith in progress is not a conviction belonging to strength, but one belonging to acquiescence and hence to weakness.
>
> (Wiener 1989 [1950])

If indeed one must contemplate human potential already ringed by socio-technical systems, the question becomes one of how to "unfold" this reality critically. Wiener was a prescient cybernetic theorist but also an insider in the US national security apparatus. One of his earliest critics was prescient too. Writing in *Le Monde* of Wiener's theories, Pere Dubarle worried,

> We may dream of the time when the *machine a gauverner* may come to supply whether for good or evil the present obvious inadequacy of the brain when the latter is concerned with the customary machinery of politics. ... In comparison with this, Hobbes' *Leviathan* was nothing but a pleasant joke. We are running the risk nowadays of a great World State, where deliberate and conscious primitive injustice may be the only possible condition for the statistical happiness of the masses: a world worse than

hell for every clear mind. Perhaps it would not be a bad idea for the teams at present creating cybernetics to add to their cadre of technicians, who have come from all horizons of science, some serious anthropologists.

(*Le Monde* 28.12.1948)

Notes

1 Salter's claim here, that airports are essentially systems "without systematizers, convergence without coordination" does not pass even the most basic of empirical tests. Every single international airport in the world is governed in whole or part by means of transposition or enactment of the regulations set out by the International Civil Aviation Organization (ICAO). Owing to this, and to innumerable other regulations, standardized processes and risk management, airports are islands in a single bureaucratic archipelago. They have, therefore, specific committees, offices and individuals responsible for systematizing and coordination, shown in job titles, for example. If one's "assemblage" theory points to a false reality, then one must ask: what's the point of "assemblage" theory?
2 Of course, to discuss "seeing" here is to talk about more than visualization—just as fingerprinting relied on the filing cabinet, contemporary visual identification requires standardized processing, databases, system search and retrieval. For example, when a traveller passes through the (increasingly common) automated border control (ABC) gates in an airport they are "seen" by the system, denoting in this case the matching of intrinsic physical characteristics to bioinformation coded and stored on a travel document. One might assume that this is a "closed" system; however, it is reliant on standardized tools and documents. Interoperability, as commonly described, refers to how increasing standardization enables searching across databases, perhaps flagging a lost or stolen passport at an ABC gate. From the perspective of a border guard, then, a closed system forms part of a larger apparatus for sorting, sifting, and controlling—seeing as multi-level bioinformation processing.
3 It is interesting here to note that Dickens's description is perfectly understandable to para-ethnographic interlocutors. The authors discussed the description with airport police as part of a facilitation session in 2017, and it formed the basis of an hour-long unpacking of physical force, the projection of authority, the psychology of crowds, etc. There is a panoply of scholarship to draw on to pursue any one of a number of themes—perhaps "human capital", perhaps even the responsible projection of force (e.g. Westbrook 2010). This description of policing is not, however, illuminated by reference to surveillance studies or assemblages (e.g. Salter 2008, Adey 2010).

References

Adey, Peter. 2010. *Ariel Life*. Oxford: Wiley-Blackwell.
Adey, Peter. 2008. Mobilities and Modulations: The Airport as a Difference Machine. In *Politics at the Airport*, edited by Mark Salter, pp. 145–160. Minneapolis: University of Minnesota Press.
Amin, Ash and Nigel Thrift. 2016. *Seeing like a City*. Cambridge: Polity.
Augé, Marc. 1995. *Non-Places: Introduction to an Anthropology of Super-modernity*. London, NY: Verso.
Bauman, Zygmunt. 2005. *Globalization: The Human Consequences*. Cambridge: Polity Press.

Brenner, Neil, David J. Madden and David Wachsmuth. 2011. Assemblage Urbanism and the Challenges of Critical Urban Theory, *City* 15 (2): 225–240.

Broome, André and Leonard Seabrooke. 2012. Seeing Like an International Organisation, *New Political Economy* 17 (1): 1–16.

Chalfin, Brenda. 2008. Sovereigns and Citizens in Close Encounter: Airport Anthropology and Customs Regimes in Neoliberal Ghana, *American Ethnologist* 35 (4): 519–538.

Coulton, Jeremy Alan. 2014. *Terminopolis or Terminal Institution? A Sociological Examination of the Institutionalised Airport Terminal.* PhD Dissertation in Sociology, University of Salford, School of Humanities, Languages & Social Sciences.

Dickens, Charles. 1851. On Duty with Inspector Fields, *Household Words.* December 6.

DiDomenica, Peter J. 2011. Statement of Detective Lieutenant Peter J. DiDomenica before the US House of Representatives Committee on Science, Space, and Technology, Subcommittee on Investigations and Oversight, *TSA SPOT Programme: A Law Enforcement Perspective*, 6 April 2011, https://science.house.gov/sites/ republicans.science.house.gov/ les/documents/hearings/2011%2003%2031%20DiDomenica%20 Testimony.pdf.

Ferguson, James. 2005. Seeing like an Oil Company: Space, Security, and Global Capital in Neoliberal Africa, *American Anthropologist* 107 (3): 377–382.

Fourcade, Marion and Kieran Healy. 2017. Seeing like a Market, *Socioeconomic Review* 15 (1): 9–29.

Gordon, Alastair. 2004. *Naked Airport: A Cultural History of the World's Most Revolutionary Structure.* Chicago: University of Chicago Press.

Haggerty, Kevin D. and Richard V. Ericson. 2000. The Surveillant Assemblage. *The British Journal of Sociology* 51 (4): 605–622. doi:10.1080/00071310020015280.

Hardt, Michael and Antonio Negri. 2003. *Empire.* Cambridge, MA: Harvard University Press.

Hill, Kashmir. 2020. The Secretive Company that might End Privacy as We Know It, *New York Times*, January 18. Available at: www.nytimes.com/2020/01/18/technology/clearview-privacy-facial-recognition.html.

Holmes, Douglas and George E.Marcus. 2005. Cultures of Expertise and the Management of Globalization: Toward the Re-functioning of Ethnography. In *Global Assemblages: Technology, Politics, and Ethics as Anthropological Problems*, edited by Aihwa Ong and Stephen J.Collier, 235–252. Oxford: Blackwell.

Jones, Calvert W. 2015. Seeing like an Autocrat: Liberal Social Engineering in an Illiberal State, *Perspectives on Politics* 13 (1): 24–41.

Latour, Bruno. 2005. *Reassembling the Social: An Introduction to Actor Network Theory.* Oxford and New York: Oxford University Press.

Lyon, David. 2008. Filtering Foes, Friends and Foes: Global Surveillance. In *Politics at the Airport*, edited by Mark Salter, pp. 29–51. Minneapolis: University of Minnesota Press.

Maguire, Mark. 2014. Counter-terrorism in European Airports. In *The Anthropology of Security: Perspectives from the Frontline of Policing, Counter-terrorism and Border Control*, eds. Mark Maguire, Catriona Frois, and Nils Zurawski, pp. 118–138. London and New York: Pluto Books.

Maguire, Mark. 2018. Policing Future Crimes. In *Bodies as Evidence: Security, Knowledge and Power*, eds. Mark Maguire, Ursula Rao and Nils Zurawski. Durham, NC: Duke University Press.

Maguire, Mark, and Fussey, Peter. 2016. Sensing Evil: Counterterrorism, Techno-science and the Cultural Reproduction of Security, *Focaal* 75 (3), 31–45.

Magnusson, Warren. 2011. *The Politics of Urbanism: Seeing like a City.* London and New York: Routledge.

Marcus, George E. and Erkan Saka. 2006. Assemblage, *Theory, Culture and Society*, 23 (2–3): 101–106.

Mathiesen, Thomas. 1997. The Viewer Society: Michel Foucault's 'Panopticon' Revisited. *Theoretical Criminology* 1 (2): 215–234. doi:10.1177/1362480697001002003.

McFarlane, Colin. 2011. Assemblage and Critical Urbanism, *City* 15 (2): 204–224.

Murphy, Eileen, and Maguire, Mark. 2015. Speed, Time and Security: Anthropological Perspectives on Automated Border Control, *Etnofoor: Anthropological Journal* 28 (1): 29–43.

Rumford, Chris. 2012. Towards a Multi-perspectival Study of Borders, *Geopolitics* 17 (4): 887–902.

Salter, Mark, ed. 2008. *Politics at the Airport*. Minneapolis: University of Minnesota Press.

Scott, James C. 1988. *Seeing Like a State: How Certain Schemes to Improve the Human Condition Have Failed*. New Haven, CT: Yale University Press.

Soeters, Joseph. 2018. *Sociology and Military Studies*. London and New York: Routledge.

Westbrook, David. 2008. *Navigators of the Contemporary: Why Ethnography Matters*. Chicago: University of Chicago Press.

Westbrook, David. 2010. *Deploying Ourselves: Islamist Violence, Globalization and the Responsible Projection of US Force*. New York: Great Barrington Books.

Wiener, Norbert. 1989[1950]. *The Human Use of Human Beings: Cybernetics and Society*. London: Free Association Books.

10 Surrender

(Bio)information in the era of the pandemic in South Korea

Kiheung Kim and Jongmi Kim

Two tales of quarantine

Scenario 1: Around the time of the end of the universal lockdown in the UK, I decided to cross the border and travel from South Korea to London to see my family. Crossing the border is not always an easy task in this pandemic situation. However, during the height of the first wave of Covid-19 in the UK, it was almost impossible to cross the border. Also, the British government began to impose quarantine rules with the exception of only some countries. Fortunately, South Korea was on the list of the quarantine exemptions. When I arrived at the airport in London, as expected the whole megastructure was nearly empty and an eerie mood filled the space. However, surprisingly, all that I needed to do was hand over my personal details, including address and phone number, and no health conditions were checked. That's it. No quarantine or any messages about careful behaviour. The border was fully opened without any safeguard.

Scenario 2: Korea had a troubling time in the early days of the pandemic. Due to its geopolitical proximity to China, when the unknown virus emerged in Wuhan, China in December 2019, everyone predicted that Korea would be the next victim of the virus. When I returned to Korea from the UK in February 2020, the airport was already controlled and travellers from outside were strictly checked and the self-quarantine app had to be downloaded. The way back home from the airport was long and winding: the airport police escorted every traveller to their next mode of transportation, including buses and trains. During the two-week quarantine period, the app I installed on my mobile phone constantly checked my movements; while a civil servant took care of my welfare, and provided boxes of foods and daily necessities. Of course, my locations and movements were controlled and monitored by the authorities, but there was no chance to raise the question as to why the state intruded upon my personal space and information. Two tests at the beginning and the end of the quarantine were conducted. After the quarantine, I returned to normal life.

Introduction

Since the mysterious pneumonic disease was reported in Wuhan, China in December 2019, South Korea has been unable to escape from the Covid-19

DOI: 10.4324/9780367810030-10

pandemic. In the early stages of the pandemic, South Korean society had an optimistic view of containing the virus effectively by using tight measures including quick testing and digital surveillance. However, on February 18, 2020, the infamous "patient 31" was confirmed as carrying Covid-19 through various neighbourhoods of Daegu city. The patient had also attended two large services at her transnational religious cult, *Shincheonji Church of Jesus*, causing the number of Covid-19 cases to spike and rise from 30 to 977 in just eight days. It became a nation-wide crisis and South Korea became one of the hot-spots of the Covid-19 pandemic. The Korean Centre for Disease Control and Prevention (KCDC) immediately decided to launch a series of containment tactics including a mass testing of suspected cases (comprehensive test of all 212,000 *Shincheonji* cult members), digital tracking methods (tracing the movement of mobile phones and credit cards) and the deployment of new testing technologies (drive-thru and walk-thru methods). The early and quick containing efforts seemed to produce positive outcomes: a decrease in the infection rate and a low mortality rate. In February 2021, the total number of confirmed cases of Covid-19 in Korea was 80,800 cases and the number of deaths sat at 1,470 (KDCA, 2021). Simultaneously the total number of British cases had reached 3.9 million and while deaths numbered some 112,000 (Public Health England 2021). Since the *Shincheonji* cult incident, there have been two more big waves of Covid-19 on the Korean peninsula. Each time, the Korean authority implemented a rapid and strict mass testing and trace of the confirmed patients' movements.

The Korean experience of containing the disease has been regarded as a successful model, while most western countries, who failed to contain the virus in the early stages, had to implement draconian lockdown measures. The three pillars of containing the disease in Korea are: test, trace and isolation. All three elements are highly dependent on tracing technologies and aggressive isolation of potential patients. The processes, as a whole, are inseparable. Once samples from potential patients are taken, the individuals' bioinformation is circulated through laboratories. After the test sample is taken, the potential patients should install an app in their mobile phone to trace movements; the app transmits detailed movements and health conditions including body temperature, cough, sore throat, breathing difficulties. The (bio)information is designed to be used only for assessing prospective patients who show symptoms of Covid-19. However, there is still the potential danger related to the disclosure of patients' private information by authorities and companies that deal with health data (Cox 2020, Zhang 2020). Although the Korean response has been widely praised as the success story of a quickly responding government's effective usage of containment technologies, some have pointed out that the Korean case is simply another example of hygienic authoritarian measures that demand the sacrifice of citizens' privacy and basic human rights.

Against this background, we ask: what social and historical factors play a part in handing over citizens' rights and privacy in an effort to contain disease? This chapter focuses on examining the social and historical factors that have led Korean society to decide to sacrifice its citizens' rights. In particular, the chapter deals with three crucial factors that have accumulated and institutionalized in Korean society: shaping democratic (bio)citizenship through a series of experiences of democratic movement since the end of 1980s; a decade-long reconstruction of disease controlling and preventing systems after experiencing recent outbreaks of zoonotic infectious diseases including SARS, MERS, Foot-and-Mouth disease and African Swine Fever; a new form of government that is obsessed with transparency and direct democratic processes following the recent experience of the anti-government movement (so-called candlelight movement) that led to the impeachment of previous president Park Geun-Hye.

Acquiescent and authoritarian mind?

Soon after the news of relatively successful containment of the disease in Asian countries like Korea, Taiwan, and Singapore began to be compared with the implementation of draconian lockdown measures in European countries in the early stage of the pandemic, some commentators tried to analyse how and why East Asian countries were quickly containing the virus without having a brutal mandatory lockdown. In particular, the focus was centred on the issue of acquiescent acceptance of the governing using personal information (Martinelli et al 2020), including bioinformation about the condition of prospective patients such as daily changes in body temperature, cough, sore throat and breathing difficulties. Furthermore, the aggressive test regime in Korea highlights the potential danger of data breaches. In one study, it was found that the contact trace data could expose sensitive private information (hobby, religion) as well as patients' social relations and workplaces (Jung, Lee, Kim and Lee 2020). As mentioned, when the *Sincheonji* cult incident caused a social unsettling in March, the most hotly debated topic was, to some extent, whether citizens' personal bioinformation should be handed over to the authority for containing the virus (Bendingfield 2020, Noh 2020). The *Sincheonji* cult is a local church which gathers secretly and aims to overtake conventional Christian churches. For this reason, it is normal for the members of the cult not to reveal their identities, and their movements were not opened to the public. However, the infamous 'patient 31' refused to reveal her movements and people with whom she had had contact, and the *Sincheonji* cult became the main target of blame (Hodge 2020, Lanese 2020). Within a week, the total number of confirmed cases spiked from 30 to 800. The majority of confirmed cases turned out to be members of the cult. The government insisted that the cult should hand over the whole list of members for comprehensive mass testing and tracing of their movements. The church rejected the demand initially but, owing to increasing public pressure, they surrendered the list of members to take Covid-19 tests. Consequently, the infection number was slowly abated.

The central issue of the debate has been closely related to how to use the personal bioinformation that is embedded in digital equipment including mobile phones. In the early days of the pandemic, the Korean health authority, the Center for Disease Control and Prevention (KCDC), launched a scheme to trace the movements of the confirmed patients. The trace system shows where the patients visited, how they moved around and who the patients contacted: in the course of figuring out the patients' movements, digital data and use of credit cards can also be accessed by the health authority. In the early stages, some personal information was exposed and many people were identified: who they were and where they worked. The second wave of infection in May was believed to have been initiated by club-goers gathered together in the *Itaewon* district of Seoul: a popular district for tourists, clubbers and foreigners. As a result of tracing their movements, some of the confirmed patients were shown to have visited gay clubs, which led to misogynic and homophobic reactions. In particular, many blamed those patients who had visited the clubs. Misogynic and homophobic backlash was quickly spreading and the gay community in Korea feared that their identities could be forced to be revealed (Kim 2020). Due to this backlash and these reactions, those who had simply passed through the *Itaewon* district also avoided having the Covid-19 test because they were afraid that personal information would be publicized. The authority quickly realized that this growing homophobic mood could jeopardize containment efforts, which had been regarded as relatively successful. The Korean CDC persuaded people to take part in testing for Covid-19 by stating that they could do so without indicating a name, an address or any personal details – except a mobile phone number. This new approach for encouraging people to get tested anonymously proved to be more effective. The so-called *Itaewon* incident was settled by the decision-making of the Korean CDC (Kang et al. 2020). The Korean authority persuaded the public by guaranteeing that personal bioinformation collected from the extensive PCR test for Covid-19 would be used only for diagnostic purposes. Soon after the MERS outbreaks in 2015, the parliament amended the Infectious Disease Prevention Act to secure transparency in the handling of health information of infectious disease patients and suspicious persons (Lim 2020). However, there is still the potential danger of misuse of the bioinformation, which could be collected, stored and processed by the government; as well as by private companies that are operating surveillance technologies and collecting biosamples for the PCR test (S-R. Kim 2020). An alarming example is that of the thermal imaging cameras used to scan people's face and temperature, which are rapidly being installed in many places including restaurants, shopping malls, school, and other public buildings. The technology can trigger an alert if a temperature above 37 degrees is identified and the information can be stored (Cox 2020).

As a repercussion of the relative success of containing the virus in Korea, debates have raged amongst western intellectuals as to why Koreans have blindly accepted the cessation of personal information. In particular, the digital trace system has been aggressively implemented without serious criticism or resistance from the public. In a survey, 68% of Korean people responded to accept the current level of information sharing even if it sacrificed individual privacy rights (Ju and You 2020). This public attitude provoked different intellectual interpretations relating to the relatively blind acceptance of government intervention in the area of personal information. It is baffling situation for some academics, particularly in western societies where popular resistance has been growing. The first and boldest explanation came from a Korean-German philosopher, Byung-cheol Han. In his short article in a Spanish newspaper, Han discusses the possibility of the return of authoritarian states (Han 2020). What he focused on was the European reactions to the virus, which tend to be to close borders and block foreign entry to the territory. This new authoritarian tendency, he warns, is the suppression of individual freedom and excessive consumption in the name of globalization. Then, he moved this discussion to cases of East Asian countries including China. Korea and Taiwan. They are relatively successful, he claimed, because those countries operated with a different type of power (Sarka 2020). The East Asian countries have deployed mass testing and digital tracing technology using mobile apps, in order to identify and trace infected persons alongside people they have contacted. During the last decade, digital surveillance technologies have been developed and deployed in every corner of these countries. The problem is, according to Han, that this digital surveillance system was possible because privacy was not a matter of concern in these societies. People have long accepted the government's intervention without raising any questions or criticism. Han's argument also resonated with French philosopher Guy Sorman (Béglè 2020). What Han and Sorman have asked is why people in East Asian countries have not felt any problem with the collection of private information including physical movements, social contacts and personal health data without proper consent. The reason is simple: the people in East Asian countries have no sense of privacy or individual freedom. Han and others find that the hierarchical status of East Asian countries is rooted in the long historical tradition of Confucianism. In other words, the reason why people in East Asia have followed the rules of the state's authority blindly is that historically shaped Confucian traditions are still working at work, playing a significant role in determining people's behaviour. This type of explanation can be easily found in many pieces of literature of that elucidate the basis of historically embedded Confucian ideology (Kasdan and Campbell 2020, Kang 2020). Their assumption is that people in East Asian countries including Korea are culturally conditioned to be deferential to authority, adhere to norms, and pursue social harmony by virtue of their Neo-Confucian traditions (Im 2014, Im, Campbell and Cha 2013, Kasdan and Campbell 2020). It means that people are collectively oriented and there is not much room for individual self. As Han highlighted, this Confucian tradition has led people to be less reluctant and more obedient than in Europe (Han 2020). Daily life is more strictly organized than in Europe, Han claims.

However, the Confucian traditions and paternalist states of East Asian countries cannot explain detailed contextual responses from different countries in this region. The claim that a few sets of cultural traditions determine people's way of behaviour may be too deterministic. Surely, in some sense, behaviour can be shaped by a long tradition. Nevertheless, it is not possible to understand people's specific behaviour on the basis of a few cultural traditions. A misconstruction of Confucian tradition has sometimes been used as a framework to disparage different behaviours and achievements of people in East Asian countries. This tendency is based upon a certain assumption that the people are regarded as subordinated and muted by authorities (Spivak 1988): an interpretation that has been a stereotypical reading of behaviour without agency. This popular image of East Asian societies has been constantly reproduced and circulated in mass media (Béglè 2020, Escobar 2020, Laurent 2020, Lu et al. 2020, Martin and Walker 2020, Maçães 2020, Wintour 2020, Zhou 2020). By highlighting the unique Confucian mentality in East Asian countries, those intellectuals including Han and Sorman reproduced the Orientalist fantasy of otherness. In other words, East Asian societies are essentially different from the West (Noh 2020: 37). The Confucian collectivism that East Asian cultures adhere to prioritizes the community over the individual and conforms to the state authority, is still controversial and unsubstantiated (Cha 2020, Choi 2020, Leonard 2020, Park 2020b, Rhode 2020, Thompson 2020). Furthermore, arguments associated with ancient Confucian influence cannot explain the subtle differences between peoples in East Asia. So-called East Asianess is often too oversimplified to embrace variations within this region. Instead, it should look at more detailed and contextualized modern experiences. In particular, Korean responses to the coronavirus need to be explained by its recent and turbulent political situation of democratization, as well as its experiences of disease.

Democratization and disease experience

As mentioned, many pieces of research and media commentaries have focused on cultural differences that lead to a different response to Covid-19. In particular, the public attitude towards the method of dealing with personal information collected by track and trace technology has caused a number of controversies. Many have attempted to find the cultural essence that explains the different attitudes. For instance, a recent study shows that the degrees of tightness and looseness of social norms played a crucial role in determining the success of containing efforts (Gelfand 2021, Gelfand et al. 2021). Although this approach looks more or less convincing, still the questions that remain, regarding detailed variations of people's attitudes towards technologies, can be reduced into a few cultural elements. It might be partly possible, but the focus needs to shift. Neither the Confucian tradition nor tightness/looseness of culture can explain specific features of responses to the handling of private information. Individual movements and behaviours as well as some health conditions can be traced by checking smartphones and credit cards. In European countries, this type of government intervention could face strong hostile resistance and become a subject of controversy (Bock et al. 2020, Cho et al. 2020, Hendl, Chung and Wild 2020, Ienca and Vayena 2020, Lucivero et al. 2020, Parker et al. 2020). So what makes people allow the authorities to monitor their movements and usage of credit cards?

The publication of the confirmed patients' contact-tracing data, by the government in South Korea, is a good example to understand the way of dealing with personal digital data. The Covid-19 containment efforts of South Korea have been regarded as relatively successful compared to other countries. The first Covid-19 case was detected on 20 January 2020 but the largescale outbreak started from the group belonging to the *Shincheonji* Church of Jesus in Daegu (Choe 2020, Park 2020a). The outbreak here led to a national crisis, and many countries blocked Korean citizens' entries across their borders and imposed a travel ban[1] to the country. The government response to Covid-19 was quick, sharp and aggressive from the early stage. Three main elements of the response are tests, contact tracing and isolation (Dudden and Marks 2020, McCurry 2020). The combative containment efforts resulted in slowing down the infection rate of the disease. While many countries struggled to contain the virus and ran out options other than mandatory lockdowns, South Korea effectively managed to contain the virus. This policy of mass testing and the intensive tracing of the movements of patients has been praised by the WHO (Keck 2020, World Health Organization 2020). At the centre of the containment efforts, digital tracing of confirmed patients played a significant role (Sonn 2020). South Korea has been known as one of the heaviest CCTV-installed countries, and that with the highest proportion of cashless transactions, in the world.[2] This heavy infrastructure in city areas has been criticized on the grounds that densely installed CCTV invades the privacy of citizens, and that the whole society is ruled with Big Brother-like authority.

The digital infrastructure of South Korea has become an accidental hero for containment efforts. The tracking system using credit cards, CCTV and the GPS of smartphones found the movements, locations and contacts of patients with others, effectively. Once the app is installed in a smart phone, daily changes in the health conditions of citizens are monitored and reported to the local health authorities. In the early stages, the intrusion of privacy became a big problem because when the authority publicized detailed movements of confirmed patients and suspected contacts, it easily exposed patients' working places and place that they had visited, with the result that many people could identify who the patients were (Jeong and Son 2020, G. O. Lee 2020). This technological surveillance system was able to invade patients' privacy, as well as becoming a powerful tool to inform people who had been around those patients recently. The possibility of overexposure of private information was reviewed and revised by the authority (Cellan-Jones 2020, H. K. Lee 2020, Zastrow 2020). The revised version of the contact-tracing system has been widely accepted by the South Korean public and even Victor Cha claims that Asia's use of digital data can be classified as "social tagging", rather than "digital surveillance" or "digital tracking". This is because "the usage of the app technology in non-autocratic societies is moving away from the general surveillance practised in China" (Cha 2020: 38).

What many commentators paid attention to is how the public in South Korea allowed the government to intrude on their private information, and where the public trust comes from. "It seems unlikely that contact-tracing methods would be acceptable in western countries, where a vigorous debate is underway about whether the contact-tracing app that stores data centrally too big a threat to privacy" (Cellan-Jones 2020). The Korean case demonstrates the significance of the concepts "democracy" and "disease experience". The former is closely associated with building public trust. Unless public trust exists in containment efforts, the cutting-edge digital technologies manipulated for contact tracing would not work. Public trust is the central element. As discussed, many intellectual depictions of East Asian societies are that they are homogeneous, obedient and contingent on a collectivist mindset. However, historical experiences show a different picture from the Orientalist descriptions. Between the years 1961 and 1997, while military dictatorship ruled the country, Korean society experienced a series of turbulent events, in the pursuit of democracy. There were countless popular movements to restore democracy against the dictatorship, which, in turn, antagonistically suppressed the civil disobedience "insurgencies". However, the popular democratic movement won out, and achieved the right to vote for an elected president in 1997 (Kim 2000, Lee 2007, Moon 2002, Oh 2012). Furthermore, since democratization in 1997, several mass protests against corrupted powers have led to construct the identity of a democratic citizenship which actively and voluntary participates in debates on social and political issues. Accumulative public distrust of government played a crucial role in impeaching the former president Geun-Hye Park in 2017 (Doucett 2017, Heo and Yun 2018, Shin and Moon 2017). One of the contributing factors in causing distrust of the government in 2017 was the failed response to the MERS (Middle East Respiratory Syndrome) outbreak in 2015. Park's administration's failed response had damaged public approval of the government. The government was to blame for lack of transparency, which was due to their act of making policy for containing the MERS outbreaks in secret. The health authority refused to disclose the facilities where patients were being treated and their movements (Gaudin 2020). This is a crucial proof that people in East Asian societies, particularly in Korea, are not naturally obedient or lacking a sense of private interests, which many western academics and commentators have inclined towards in interpreting the Orientalist point of view.

There are two meaningful transformations: the first is the experience of democratization as a driving force to demand more transparent implementation of policy. One of the reasons for impeaching President Park in 2016/17 was a corruption charge including secrecy in her administration. After the so-called candlelight demonstration that led to the impeachment, the public's pressure upon the newly elected Moon Jae-in's administration has been powerful enough to be the driving force in implementing a new type of policy. Correspondingly, the Moon administration has not been able to ignore the popular insistence and has tended to be obsessed with the transparency of policy (Gaudin 2020). Another transformation is the experience of disease. Since the new millennium, there have been waves of outbreaks of infectious diseases in animals and humans. The Covid-19 pandemic has been framed in the context of the 2003 Severe Acute Respiratory Syndrome (SARS), 2009 H1N1 influenza in humans,

2010 Foot-and-Mouth Disease (FMD) and Avian Influenza (AI) in animals and 2015 MERS outbreaks. In particular, SARS and MERS were caused by the same type of coronaviruses. The concept of "disease experience" plays an important role in understanding the full anatomy of governance of disease. As Hinchliffe, Bingham, Allen and Carter claimed, outbreaks of diseases are not simple events – rather, they are closely related to various spatial and temporal orderings of the outbreaks – and suggested a new concept, "disease situation" (Hinchliffe et al. 2017: 52). In this sense, disease experiences in the disease situation can be a useful framework to understand the whole context of the outbreaks.

Wave after wave the public suffered, and mistakes made in the governance of the diseases were corrected and revised. During the course of the disasters, disease experiences transformed whole attitudes towards the way of containing diseases. Examples can easily be found: in Taiwan, the experience of the SARS outbreak resulted in the transformations of health emergency planning such as the National Health Command Center and Central Epidemic Command Center (Wang, Ng and Brook 2020, Cha 2020). Similarly, the institutional and technological transformations in South Korea are radical and comprehensive. In the case of Korea, a major problem was that, before the outbreaks of infectious diseases, there was no proper control centre for governing diseases. Prior to the outbreak of SARS in 2003, biopolitical governance of diseases and population are not recognizable in the context of South Korea. However, the SARS outbreak alarmed policymakers and led them to establish the Korean Centers for Disease Control and Prevention (KCDC) in 2004, though the organization still belonged to the Department of Health and Welfare. In 2015, an older coronavirus outbreak, MERS, became a pivotal moment in reforming the whole infrastructure of controlling diseases (Fung et al. 2015, Kim 2016, 2020).

The MERS outbreak sent big shockwaves through Korean society. It began with one person who had travelled from Saudi Arabia, and resulted in 186 confirmed cases with 38 deaths. Many criticized the government's handling of the disease due to its secretive nature (Park and Chung 2021). Although enormous popular pressure to publicize the information of where infection occurred, as well as the movements of patients, was growing, the conservative Park administration failed to unveil the necessary information. Consequently, the public established several digital apps to provide information on the list of hospitals treating positive cases, and movements of patients, voluntarily (Tynan 2020). Finally, in response to the criticisms, South Korea amended its "Infectious Disease Control and Prevention Act" to expand the authority of the health minister to ensure the citizen's "right to know" the path of viral infection. This amendment has allowed health authorities to collect extensive personal data from the mobile phone, credit cards, GPS information and social network system (SNS) information in the case of a public health emergency (Landman 2020). Furthermore, in 2020, the newly elected Moon Jae-in's administration decided to expand the role of KCDC establishing a bigger organization called "Korea Disease Control and Prevention Agency, KDCA" for the control and governance of infectious diseases. The establishment of a new agency and amending rules needed public support and trust.

Without public trust, the whole system of testing and tracing could not have worked properly. Interestingly, the Korean public's opinions about surveillance have been overwhelmingly positive (Realmeter 2020). It is always hard to strike the balance between the public's right to know and the individual's right to privacy. On the governmental level, epidemic preparedness was recognized as a core biopolitical element after several major outbreaks of infectious diseases in humans and animals. On the other hand, in the public domain, the public needs to be assured that the governmental policy is being implemented in a transparent and democratic way. During the early period of the pandemic when most countries had mandatory lockdown, local businesses and organizations voluntarily took necessary actions, while schools and workplaces made the swift transition to online classes and work at home. Access to private (bio)information has offered up voluntarily to the health authorities, who consequently should provide a firm promise that their intervention will be implemented in a transparent and democratic manner. It is interesting to see how the Korean public could accept invasive technologies as part of a social contract between the state and society in a time of crisis, rather than rejecting them. This type of social contract is not so much associated with ancient rules including the Confucian tradition. Rather in the course of experiencing disastrous outbreaks of infectious diseases and a turbulent democratization process, a co-construction of government's responsibilities to provide the best information possible to the public and the civic obligation to cooperate to reduce the infection has emerged.

Coda: limitations and problems

Inevitably, it is hard to disregard the long-standing dichotomy between the East and West. In other words, from the western point of view, there should be a "modern self (European west) and premodern others" or "democratic individualism and Confucian collectivism". However, since independence from Imperial Japan, Korea has followed a rapid pathway of modernization and postcolonial development. The intertwined nature of postcolonial development cannot be explained by the dichotomies. It is necessary to bring in the collective experience of more recent, turbulent years. In the context of the Covid-19 pandemic, South Korea's relative successful in "flattening the curve", using aggressive test and tracing strategies, has baffled many western commentators and intellectuals. Their explanations are still based on what Bruno Latour called "the Great Divides" of modernity (Latour 1993). Within this, their main argument is that the long-standing Confucian tradition is the key element in allowing the public to uncritically accept the state's access to personal digital data, as discussed.

However, what this research found is that voluntary cessionary behaviour does not stem from the ancient collectivist mentality. Instead, recent experiences of democratization and outbreaks of infectious diseases provide more comprehensive understandings of the current pandemic situation in South Korea. Nevertheless, a question is still to be answered: can the state-centred Covid-19 containment efforts be a pathway to a new type of authoritarian regime? The voluntary civil engagement in the containing of Covid-19 led to an overwhelmingly positive opinion of the Moon administration. As a result, the public supported Moon's Covid-19 policies and the ruling Democratic Party had a landslide victory in the general election of April 2020. This landslide victory gave the Moon administration an opportunity to implement policies to contain the disease. However, the state's intervention and accessibility of personal data are sometimes implemented excessively. Although the health authority has tried to revise the methods of managing personal data constantly, digital data of confirmed patients has been identified and exposed easily on the internet. For instance, when movements and credit card records of two patients were publicized in February, people could find the fact that they were a couple engaged in an extramarital affair (Fendos 2020). Furthermore, the chair of the National Human Rights Commission of Korea expressed concern that "excessive disclosure of private information could cause people with symptoms to avoid testing" (Zastrow 2020).

As the Moon administration was elected directly after the mass demonstrations against his predecessor, Park Geun-Hye, and her subsequent impeachment in 2016–2017, the direct democratic movement has been a fundamental underpinning of the power of the current government. This tendency towards direct participation sometimes creates a paradoxical situation regarding democratic processes. In the case of the incident of the *Shincheonji* Church in February, over 1.4 million people took part in a petition to demand that the central government "forcefully disband the Church" (Gaudin 2020). The danger is that the popular mood against a cult group, or any other minority, can easily be associated with a discriminatory and anti-democratic mood in society; which is exactly opposite of what the current government outlines, that is, more human rights, social welfare and transparency. As Gaudin observes, "the counterpoise surges from direct democracy itself" (Gaudin 2020). Furthermore, after the landslide victory of the general election, the government pushed through a series of policies which unilaterally suppressed opposition voices. For instance, when Korean society found that another flare-up of infections was linked to a Christian Church-based anti-government group in October 2020, the Korean government banned more than 100 anti-government demonstrations on the National Foundation Day, sealing off every main avenue of a central Seoul square with parked police buses. This government reaction was heavily criticized, and many suspected that this excessive response was aimed at silencing voices critical of the government (Kang 2020, Shin 2020).

Despite all the concerns and limitations, the Korean public's embrace of digital surveillance has resulted in open businesses and a relatively normal life without compulsory lockdown. Also, containment efforts have been comparatively successful. Some insist that this acceptance of authority-controlled digital surveillance could pose the threat of becoming a new form of authoritarian regime, and the passive attitude of the public can only be explained by long-standing Confucian traditions. However, as seen, the acceptance of digital surveillance is a form of a voluntary social contract between the public and the authority. The way of implementing safeguards by track and trace technologies is, to some extent, invading individual privacy, but as a means of protecting themselves from infections as devastating as the Covid-19 virus, citizens can give up their part of privacy temporarily.

In the Korean context, there are two pivotal collective experiences that have transformed the mentality of the public as a whole. The first is the achievement of democracy in 1997. The experience of restoring formal democracy allowed the build-up of the concept of "democratic citizenship", which was not encouraged during the military regime.[3] More recently, the active involvement in forcing a change of governments, including impeaching the previous administration, transformed Korean society into one where people take part in democratic decision-making voluntarily. The second pivotal transformative experience is the disease experience that awakened people to realize the right to a healthy life. During the popular demonstration against importing American beef during the BSE crisis in 2008, non-governmental organizations and general citizens demanded the right to protect their healthy life, when the conservative government decided to import American beef that included specified risk materials (SRM) (Kim 2009). Subsequently, a series of outbreaks of infectious diseases like SARS, MERS, AI and FMD alarmed the public to the right of protection of citizens' lives from infectious disease, and became a main priority of the government. The disease experiences contributed to shaping a new type of citizenship that is closely associated with life and death. This is how some researchers, influenced by Michel Foucault's biopolitics, developed the concept of biocitizenship (Cooter 2010, Kim 2012, Petryna 2004, Rose and Novas 2007). The formation of democratic citizenship, as well as biocitizenship, in South Korea, may be playing a pivotal role in transforming attitudes towards voluntary participation in containment efforts during the Covid-19 pandemic. The voluntary cession of personal information should be understood in the context of detailed collective experiences rather than being reduced to a few elements, like Confucianism.

Epilogue: personal refection

Recently, again, I crossed the border between the UK and South Korea during the third British lockdown and travel restrictions from/to the UK in Korea were even stricter, as the new variant of Covid-19 was prevalent. Within 15 days, including the 14-day quarantine in Korea, I had to go through PCR testing five times, as well as a 24-hour lock-up in a quarantine facility (the first test was part of the release to travel scheme in the UK, followed by two compulsory tests before and after the quarantine, along with one in the middle). As soon as I arrived in the airport in Korea, the local authority, where I live, was alerted to the fact that I had flown in from the UK. Since the emergence of the new variant in the south of England, all of the direct flights between the UK and Korea were cancelled; and overseas arrivals from the UK, South Africa and Brazil have become specially managed subjects. The local authority had power to implement an administrative order to confine those people in a government facility, even if the person is not infected. The tough and rigorous way of restricting individual freedoms in the name of protecting the public's safety has an ambivalent nature: uncomfortable feelings that private freedom is being compromised, as well as comfortable feelings of being protected from the disastrous virus. One thing is certain: that this contradictory state must be ended as soon as the pandemic is abated. Moreover, the current legislation should be amended to find more justifiable grounds for restricting people's privacy, in a state of emergency, by the independent governing body.

Notes

1 At the time, 187 countries have imposed travel restrictions from/to South Korea (Choe 2020).
2 According to national statistics, over 1.1 million CCTVs are installed in public buildings and places in 2019 (National Statistics Korea 2019). In the private area, CCTV cameras have been distributed and everyone can be captured an average of 83.1 times per day and every nine seconds while travelling (Sonn 2020).
3 During the military regime, the nationalistic mentality had been promoted and always outlined a concept of national identity instead of citizenship. The formation of citizenship and civil society has been oppressed or militarized by the regime. However, since the democratization in 1997, the desire to build citizenship has exploded (Moon 2005).

Reference

Béglè, Jérôme (2020) 'Guy Sorman: « Le confinement nous fait découvrir qu'avant, ce n'était pas si mal »', *Le Point* (27 April 2020), www.lepcint.fr/debats/guy-sorman-le-confinem ent-nous-fait-decouvrir-qu-avant-ce-n-etait-pas-si-mal-27-04-2020-2373021_2.php.

Bendingfield, Will (2020) 'What the world can learn from South Korea's coronavirus strategy', *Wired* (21 March 2020) www.wired.co.uk/article/south-korea-coronavirus.

Bock, Kirsten, Christian R. Kühne, Rainer Mühlhoff, Měto R. Ost, Jörg Pohle, & Rainer Rehak (2020) Data Protection Impact Assessment for the Corona App (29 April 2020). Available at SSRN: https://ssrn.com/abstract=3588172 or http://dx.doi. org/10.2139/ssrn.3588172.

Cellan-Jones, Rory (2020) 'Tech Tent: Can we learn about coronavirus-tracing from South Korea?' *BBC News* (15 May 2020), www.bbc.co.uk/news/technology-52681464.

Cha, Vitor (2020) 'Asia's Covid-19 lessons for the West: Public goods, privacy, and social tagging', *The Washington Quarterly* 43 (2): 33–50.

Cho, Hyinghoon, Daphne Ippolito & Yun W. Yu (2020) 'Contact tracing mobile apps for Covid-19: privacy considerations and related trade-offs', *arXiv*, Cornell University. https://arxiv.org/pdf/2003.11511.pdf.

Choe, Sang-Hun (2020) 'Shadowy church is at center of coronavirus outbreak in South Korea', *New York Times* (21 February 2020),www.nytimes.com/2020/02/21/world/a sia/south-korea-coronavirus-shincheonji.html.

Choi, Yon Jung (2020) 'The power of collaborative governance: The case of South Korea responding to Covid-19 pandemic', *World Medical and Health Policy* 12 (4): 430–442.

Cooter, Roger (2010) 'Cracking biopower', *History of the Human Sciences* 23 (2): 109–128.

Cox, David (2020) 'Alarm bells ring for patient data and privacy in the Covid-19 gold-rush', *The BMJ* 369: m1925.

Doucett, Jamie (2017) 'The occult of personality: Korea's candlelight protests and the impeachment of Park Geun-Hye', *The Journal of Asian Studies* 76 (4): 851–860.

Dudden, Alexis & Andrew Marks (2020) 'South Korea took rapid, intrusive measures against Covid-19 and they worked', *The Guardian* (20 March 2020), www.theguardian.com/commentisfree/2020/mar/20/south-korea-rapid-intrusive-measures-covid-19.

Escobar, Pete (2020) 'Confucius is winning the Covid-19 war', *Asian Times* (13 April 2020), https://asiatimes.com/2020/04/confucius-is-winning-the-covid-19-war.

Farrer, Martin (2020) 'Coronavirus: South Korea cluster drives huge rise in cases', *The Guardian* (22 February 2020), www.theguardian.com/world/2020/feb/22/coronavirus-south-korea-sees-huge-jump-cases-china-hubei-wuhan-outbreak-.

Fendos, Justin (2020) 'How surveillance technology powered South Korea's Covid-19 response', *TechStream* (29 April 2020), www.brookings.edu/techstream/how-surveillance-technology-powered-south-koreas-covid-19-response.

Fung, Isaac C-H., Zion T.H.Tse, Benedict S.B.Chan & King-Wa Fu (2015) 'Middle East respiratory syndrome in the Republic of Korea: Transparency and communication are key', *WPSAR* 6 (3), doi:10.5365/wpsar.2015.6.2.011.

Gaudin, Christophe (2020) 'Korean democracy in times of coronavirus', *Lettre du Centre Asie* 80 (1 April 2020).

Gelfand, Michele J.*et al.* (2021) 'The relationship between cultural tightness-looseness and Covid-19 cases and deaths: A global analysis', *The Lancet Planetary Health*, doi:10.1016/S2542-5196(20)30301–30306.

Gelfand, Michele J. (2021) 'Why countries with loose, rule-breaking cultures have been hit harder by Covid', *The Guardian* (1 February 2021), www.theguardian.com/world/commentisfree/2021/feb/01/loose-rule-breaking-culture-covid-deaths-societies-pandemic?utm_source=dlvr.it&utm_medium=facebook&fbclid=IwAR2uxvkLlidxlQjQ-lS02J8ezgUrwailVpfCpSUQPWqc-gHS2KPLLN8u3gk.

Hendl, Tereza, Ryoa Chung & Verina Wild (2020) 'Pandemic surveillance and racialized subpopulations: Mitigating vulnerabilities in Covid-19 apps', *Bioethical Inquiry*, doi:10.1007/s11673-020-10034-7.

Han, Byung-Chul (2020) 'The viral emergenc(e/y) and the world of tomorrow', https://pianolaconalbedrio.wordpress.com/2020/03/29/the-viral-emergence-y-and-the-world-of-tomorrow-byun-chul-han.

Heo, Uk & Seongyi Yun (2018) 'South Korea 2017', *Asian Survey* 58 (1): 65–72.

Hodge, Mark (2020) 'Spiral of death: How chilling story of South Korea's patient 31 is a deadly warning for Brits to take NHS track and trace app seriously', *The Sun* (6 May 2020), www.thesun.co.uk/news/11553406/south-korea-patient-31-nhs-track-trace-app.

Im, Tobin (2014) 'Buraucracy in three different worlds: The assumptions of failed public sector reforms in Korea', *Public Organization Review* 14 (4): 577–596.

Im, Tobin, Jesse W.Campbell & Seyeong Cha (2013) 'Revisiting Confucian bureaucracy: Roots of the Korean government's culture and competitiveness', *Public Administration and Development* 33 (4): 286–296.

Ienca, Marcello & Effy Vayena (2020) 'On the responsible use of digital data to tackle the Covid-19 pandemic', *Nature Medicine* 26: 458–464.

Jeong, Jonggu & Junggoo Son (2020) 'Legal analysis of Covid-19 disclosure', *Kyungbook National University Law Journal* 70: 103–131.

Ju, Youngkee & Myoungsoon You (2020) 'The outrage effect of personal stake, dread and moral nature on fine dust risk perception moderated by media use', *Health Communication*, doi:10.1080/10410236.2020.1723046.

Jung, Gyuwon, Hyunsoo Lee, AukKim & Uichin Lee (2020) 'Too much information: Assessing privacy risks of contact data disclosure on people with Covid-19 in South Korea', *Frontiers in Public Health* 8 (305). doi:10.3389/pubh.2020.00305.

Kasdan, David Oliver & Jesse W. Campbell (2020) 'Dataveillant collectivism and the coronavirus in Korea: Values, biases, and socio-cultural foundations of containment efforts', *Administrative Theory & Praxis* 42 (4): 604–613.

Kang, Jaeho (2020) 'The media spectacle of a techno-city: Covid-19 and the South Korean experience of the state of emergency', *The Journal of Asian Studies* 79 (3): 589–598.

Kang, Tae-jun (2020) 'South Korea baffled by stop-and-search, police bus walls during protests', *The Diplomat* (5 October 2020), https://thediplomat.com/2020/10/south-korea-baffled-by-stop-and-search-police-bus-walls-during-protests.

Kang, Cho Ryok*et al.* (2020) 'Coronavirus disease exposure and spread from nightclubs, South Korea', *Emerging Infectious Diseases* 26 (10): 2499–2501.

KDCA (2021) 'Coronavirus Disease-19, Republic of Korea', *Korea Disease Control and Prevention Agency* (7 February 2021), http://ncov.mohw.go.kr/en.

Keck, Frédéric (2020) 'Asian tigers and the Chinese dragon: Competition and collaboration between sentinels of pandemics from SARS to Covid-19', *Centaurus* 62: 311–320.

Kim, Sunhyuk (2000) *The Politics of Democratization in Korea: The Role of Civil Society* (Pittsburgh, PA: University of Pittsburgh Press).

Kim, Sun-Ryang (2020) 'A study on the collection of information on the patients of infectious diseases, etc. and persons suspected of contracting infectious diseases, disclosure of information on the patients of infectious diseases and protection of personal information', *Journal of Media Law, Ethics and Policy Research* 19 (3): 1–31.

Kim, Kiheung (2009) *The Mad Cow Controversy* (Seoul: Haenamu).

Kim, Kiheung (2012) 'The state-initiated development of bioscience and the emergence of biocitizenship in East Asian countries', *Asia Review* 2 (2): 43–63.

Kim, Kiheung (2016) 'Uncertainties of international standards in the MERS CoV outbreak in Korea: Multiplicity of uncertainties', *ECO* 20 (1): 317–351.

Kim, Kiheung (2020) 'The disease governing system and co-construction of human and animal diseases in South Korea' in The Human-Animal Research Network (ed.), *Relations and Boundaries-Human and Animals in the Age of the Covid-19 Pandemic* (Seoul: Podobat Publisher): 159–184.

Kim, Nemo (2020) 'Anti-gay backlash feared in South Korea after coronavirus media reports', *The Guardian* (8 May 2020), www.theguardian.com/world/2020/may/08/anti-gay-backlash-feared-in-south-korea-after-coronavirus-media-reports.

Landman, Karen (2020) 'What we can learn from South Korea's coronavirus response', *Elemental* (1 June 2020), https://elemental.medium.com/what-we-can-learn-from-south-koreas-coronavirus-response-97a4db5c9fef.

Lanese, Nicoletta (2020) 'Superspreader in South Korea infects nearly 40 people with coronavirus', *Live Science* (23 February 2020), www.livescience.com/coronavirus-superspreader-south-korea-church.html.

Latour, Bruno (1993) *We Have Never Been Modern* (Cambridge, MA: Harvard University Press).

Laurent, Lionel (2020) 'Don't ignore the good news on Covid-19 from Asia', *Bloomberg* (30 October 2020), www.bloombergquint.com/gadfly/don-t-ignore-the-good-news-on-covid-19-from-asia.

Leonard, Andrew (2020) 'Taiwan is beating the coronavirus. Can the US do the same?', *WIRED* (18 March 2020), www.wired.com/story/taiwan-is-beating-the-coronavirus-can-the-us-do-the-same.

Lee, Geun Oak (2020) 'The legal harmony between personal information protection and public interest under the Official Information Disclosure Act: Focusing on privacy infringement under the Covid-19 pandemic', *Korean Journal of Communication & Information* 103: 145–176.

Lee, Hakyung Kate (2020) 'South Korea's contact tracing shed light on extensive efforts to slow spread of Covid-19', *ABC News* (9 December 2020), https://abcnews.go.com/International/south-koreas-contact-tracers-struggle-slow-spread-covid/story?id=74621480.

Lee, Namhee (2007) *The Making of Minjung: Democracy and the Politics of Representation in South Korea* (Ithaca, NY: Cornell University Press).

Lim, Gyeo-Cheol (2020) 'Critical discussions on the use of personal information in implementing coronavirus overcoming measures', *Chungnam Law Review* 31 (4): 71–112.

Lu, Li, Srinivas Lankala, Yuan Gong, Xuefeng Feng & Briankle G. Chang (2020) 'Forum: Covid-19 dispatches', *Cultural Studies⬚Critical Methodologies*, doi:10.1177/1532708620953190.

Lucivero, Federica, Nina Hallowell, Stephanie Johnson, Barbara Prainsack, Gabrielle Samuel & Tamar Sharon (2020) 'Covid-19 and contact tracing app: Ethical challenges for a social experiment on a global scale', *Bioethical Inquiry*, doi:10.1007/s11673-020-10016-9.

Maçães, Bruno (2020) 'Coronavirus and the clash of civilizations', *National Review* (10 March 2020), www.nationalreview.com/2020/03/coronavirus-and-the-clash-of-civilizations.

Martin, Timothy W. & Marcus Walker (2020) 'East vs. West: Coronavirus fight tests divergent strategies', *The Wall Street Journal* (13 March 2020) www.wsj.com/articles/east-vs-west-coronavirus-fight-tests-divergent-strategies-11584110308.

Martinelli, Lucia, Vanja Kopilas, Matijaz Vidmar, Ciara Heavin, Helena Machado, Zoran Todorovic, Norbert Buzas, Mirjam Pot, Barbara Prainsack & Srecko Gajovic (2020) 'Face masks during the Covid-19 pandemic: A simple protection tool with many meanings', *Frontiers in Public Health* 8, doi:10.3389/fpubh.2020.606635.

McCurry, Justin (2020) 'Test, trace, contain: How South Korea flattened its coronavirus curve', *The Guardian* (23 April 2020), www.theguardian.com/world/2020/apr/23/test-trace-contain-how-south-korea-flattened-its-coronavirus-curve.

Moon, Seungsook (2002) 'Carving out space: Civil society and the women's movement in South Korea', *The Journal of Asian Studies* 61 (2): 473–500.

Moon, Seungsook (2005) *Militarized Modernity and Gendered Citizenship in South Korea* (Durham, NC: Duke University Press).

National Statistics Korea (2019) 'Statistics of number of CCTV in public buildings and places', *K-indicator, National Statistics Korea*, www.index.go.kr/potal/main/EachDtlPageDetail.do?idx_cd=2855.

Noh, Minjung (2020) 'Understanding South Korea's religious landscape, patient 31 and the Covid-19 exceptionalism', in David Kenley (ed.), *Teaching about Asia in a Time of Pandemic* (Ann Arbor, MI: Association for Asian Studies): 31–40.

Oh, Jennifer S. (2012) 'Strong state and strong civil society in contemporary South Korea: Challenges to democratic governance', *Asian Survey* 52 (3): 528–549.

Park, June & Eunbin Chung (2021) 'Learning from past pandemic governance: Early response and public-private partnerships in testing of Covid-19 in South Korea', *World Development* 137, doi:10.1016/j.worlddev.2020.105198.

Park, Nathan (2020a) 'Cults and conservatives spread coronavirus in South Korea', *Foreign Policy* (27 February 2020), https://foreignpolicy.com/2020/02/27/coronavirus-south-korea-cults-conservatives-china.

Park, Nathan (2020b) 'Confucianism isn't helping beat the coronavirus', *Foreign Policy* (2 April 2020), https://foreignpolicy.com/2020/04/02/confucianism-south-korea-coronavirus-testing-cultural-trope-orientalism.

Parker, Michael J., Christopher Fraser, Lucie Abeler-Dorner & David Bonsall (2020) 'Ethics of instantaneous contact tracing using mobile phone apps in the control of the Covid-19 pandemic', *Journal of Medical Ethics* 46: 427–431.

Petryna, Adriana (2004) 'Biological citizenship: The science and politics of Chernobyl-exposed populations', *Osiris* 19: 250–265.

Public Health England (2021) 'Coronavirus in the UK', *Public Health England* (7 February 2021), https://coronavirus.data.gov.uk.

Realmeter (2020) 'Public opinion for agreement of the compulsory Investigation in the case of a public health emergency', *Realmeter* (26 February 2020),

Rhode, Benjamin (2020) 'Living and dying nations and the age of Covid-19', *Survival* 62 (5): 223–234.

Rose, Nikolas & Carlos Novas (2007) 'Biological citizenship', in Aihwa Ong & Stephen J. Collier (eds), *Global Assemblages: Technology, Politics and Ethics as Anthropological Problems* (Oxford: Blackwell): 439–463.

Sarka, Swagato (2020) 'Pandemic and the state', *Journal of Social and Economic Development*, doi:10.1007/s40847-020-00129-7.

Shin, Gi-Wook & Rennie J. Moon (2017) 'South Korea after impeachment', *Journal of Democracy* 28 (4): 117–131.

Shin, Hyonhee (2020) 'South Korea police set up bus walls to prevent protests amid Covid-19 concerns', *Yahoo! News* (3 October 2020), https://uk.news.yahoo.com/south-korea-police-set-bus-092802193.html.

Sonn, Jung Won (2020) 'Coronavirus: South Korea's success in controlling disease is due to its acceptace of surveillance', *The Conversation* (20 March 2020), https://theconversation.com/coronavirus-south-koreas-success-in-controlling-disease-is-due-to-its-acceptance-of-surveillance-134068.

Spivak, Gayatri Chakravorty (1988) 'Can the subaltern speak?' in Cary Nelson & Lawrence Grossberg (eds), *Marxism and the Interpretation of Culture* (London: Macmillan): 271–313.

Thompson, Derek (2020) 'What's behind South Korea's Covid-19 exceptionalism', *The Atlantic* (6 May 2020), www.theatlantic.com/ideas/archive/2020/05/whats-south-korea s-secret/611215.

Tynan, Caroline (2020) 'State responses to Covid-19: South Korea, Taiwan and the power of strong democracies', *Global Policy* (4 June 2020), www.globalpolicyjournal.com/blog/04/06/2020/state-responses-covid-19-south-korea-taiwan-and-power-strong-democracies.

Wang, C. Jason, Chun Y.Ng & Robert H. Brook (2020) 'Response to Covid-19 in Taiwan: Big data analytics, new technology and proactive testing', *JAMA* 323 (14): 1341–1342.

Wintour, Patrick (2020) 'Coronavirus: Who will be winners and losers in new world order?', *The Guardian* (11 April 2020), www.theguardian.com/world/2020/apr/11/coronavirus-who-will-be-winners-and-losers-in-new-world-order.

World Health Organization (2020) 'WHO Director-General's opening remarks at the media briefing on Covid-19', *Director-General Speech of World Health Organization* (18 March 2020), www.who.int/director-general/speeches/detail/who-director-general-s-opening-remarks-at-the-media-briefing-on-covid-19—18-march-2020.

Zastrow, Mark (2020) 'South Korea is reporting intimate details of Covid-19 cases: Has it helped?', *Nature* (18 March 2020), doi:10.1038/d41586-020-00740-y.

Zhang, Hongyu (2020) 'With coronavirus containment efforts, what are the privacy rights of patients?', *The Conversation* (14 March 2020), https://theconversation.com/with-coronavirus-containment-efforts-what-are-the-privacy-rights-of-patients-131752.

Zhou, Christina (2020) 'Why are Western countries being hit harder than East Asian countries by coronavirus?', *ABC News* (24 April 2020), www.abc.net.au/news/2020-04-24/coronavirus-response-in-china-south-korea-italy-uk-us-singapore/12158504.

Afterword

Data, life and worlds in an anthropology of bioinformation

Noah Tamarkin

We might understand bioinformation as the distillation and dissemination of information from biological organisms. But through the essays in this book, we learn that it is also so much more: bioinformation is a translation, a representation, and a product. Most importantly, as this book's introduction makes clear, bioinformation is a promise reaching towards desired futures; it is thus a condensation and projection of differentiated definitions of life and its governance.

Questions of course arise: from whom, under what circumstances, and through which desires do data derive in the first place? Furthermore, what desires are then prompted by bioinformation in its many forms, and whose projects do they advance? In other words, who wants particular forms of bioinformation, and what do those forms of bioinformation in turn prompt us to want? Finally, what does the gathering, organization, and communication of biological data do?

These questions point towards the openings provided by an anthropology of bioinformation, where new meanings forged through contingent relationalities can be approached, interrogated, and narrated. An anthropology of bioinformation gets at the messiness that precedes, constitutes, and exceeds the clean lines that bioinformation promises, be they barcodes, surveys, medical devices, or databases. It investigates how standardized forms of organizing data come about and sometimes take on lives on their own. Methodologically, anthropology is especially well equipped to ethnographically examine what particular examples of bioinformation bring into focus, and what they obscure. From attention to database decisions made collaboratively in boardrooms (Van Allen, this volume) to how workers' bodies, fitted with cameras, document with visual precision the risks of e-waste's toxic exposures (Little, this volume), the essays in this book provide a blueprint for how one might do ethnography of bioinformation and a roadmap for the kinds of insights an ethnography of bioinformation can provoke. My spatial metaphors here are not accidental: these approaches to bioinformation are carefully routed and deeply grounded.

Attentive to process, power, and relationality, the anthropology of bioinformation presented here crystalizes the formations, transformations, and frictions of data worldings (Tsing 2005, Zhan 2009: 7). I therefore want to highlight how the

DOI: 10.4324/9780367810030-11

essays in this book engage with these three key aspects of anthropological approaches to bioinformation: process, power, and relationality. These are especially critical concepts through which to engage bioinformation because they allow us to get at precisely why bioinformation is such a fruitful object of research in this moment, and why we need the kinds of critical approaches presented here.

Process

Ethnographic attention to the processes through which bioinformation is produced attune us first and foremost to the question of labor. Labor here signals the work and effort required to render information from life, a process that is especially clear in Tahani Nadim's essay "All the data creatures," Adrian Van Allen's essay "Capturing genomes: the friction and flow of bioinformation at the Smithsonian," and Anisha Chadha's essay "Bioinformation in formation: inventing medical devices" (this volume). Ethnographically studying things like the labor involved in creating links between animal tissue and species data (Van Allan, this volume) or data itself as an ethnographic object that also must be understood as an ethnographic moment (Nadim, this volume) we see the impossibility of the kind of project of standardized and totalized representation of life that projects like biobanking or species barcoding proport to undertake. When thinking about these kinds of cataloging projects, we might be tempted to assess bioinformation as decontextualized data, but both Nadim and Van Allen instead show how bioinformation resists such decontextualization, carrying forward not only its provenance and conditions of production, but also other, earlier, forms of cataloging life. Interestingly, Chadha gives us another, especially visceral, way to conceptualize this same kind of resistance to decontextualization: here Indian medtech workers elicit data from bodies for medical device prototypes, which, if they go on to become mass produced, will scale up the necessarily non-standard bio-measurements of test subjects. Chadha's analysis demands that we recognize the traces of embodied specificity that then go on to inform new body/data interfaces and outputs.

Labor also becomes significant when bioinformation is itself a form of surveilling or otherwise documenting laboring bodies, whether the paid work of e-waste laborers (Little, this volume), the unpaid work of voluntary self-surveillors who hope that their data might curb the spread of Covid-19 through contact tracing (Kim and Kim, this volume), or the specialized work of airport security officers who struggle to sync multiple and incompatible biometric systems, not to mention the illicit work of airport shoplifters who seek to avoid these security officers' detection, and the paraethnographic work of anthropologists whose labor alongside security workers necessarily contributes to securitization projects that aren't their own (Murphy and Maguire, this volume). This approach to laboring bodies further challenges the idea that aspirations of totalizing knowledge are realizable by pointing out moments where intent, execution, and implication simply do not line up. It also highlights a disjuncture between the production and the usefulness of bioinformation: as multiple authors here point out, the impulse to produce more and more data can elide the question of what, exactly, the data is for (see especially Nadim and Little, both this volume).

Attention to process also highlights two other necessary components of bioinformation: data transfer and data capture. Indeed, the question of transfer opens this volume: as Gonzalez-Polledo and Posocco write, bioinformation first and foremost "marks transitions from body to data, biological substance to information, and archives to datasets" (Chapter 1, this volume). In this way, we can view data capture as itself a form of transfer; furthermore, the conditions of data collection and the aspirations for what a collection of data might do can tell us quite a bit about the politics of specific bioinformation repositories. For example, the shifting methods of a still-ongoing British National Survey of Health and Development birth cohort study sheds light on how post-World War II British anxieties about fertility and class have shifted to new concerns about aging and disease (Cruz, Tinkler, and Fenton, this volume), and the aspirations of contemporary biorepository designers for complete inventories of life resonate with those that animated colonial and imperial taxonomists' projects of earlier eras (Nadim this volume; Van Allen, this volume).

Power

Bioinformation can crystalize forms of power just as much as it can obscure them: this is why labor, transfer, and capture, discussed above, are such important foci for an ethnographic lens that seeks to understand how power is amassed and contested (see especially Van Allen, this volume). Practices of data mining are one important way that this comes into focus since this is a form of data transfer that is best understood as extraction (Gonzalez-Polledo and Posocco, this volume). As a research practice, data mining can be illuminating, as for example in Resto Cruz, Penny Tinkler and Laura Fenton's method of scavenging data to read new forms of familial power relations and participant subjectivities from the marginalia of surveys where researchers and participants recorded extraneous details not cultivated by the study design and the questions that were originally posed (Cruz, Tinkler and Fenton, this volume). However, it can also attune us to forms of exploitation and entrenchment of existing inequalities, something that Gonzalez-Polledo and Posocco emphasize in this volume's introduction. They explain that bioinformation is constitutive of rather than external to governance and sovereignty, and that this process often further marginalizes the already marginalized (Chapter 1).

We therefore need to ask, whether we are talking about forms of amassing data or mining data already amassed, not only who wants this data and how is the data gathered and stored, but also how is data rendered trustworthy and for whom; what presents and futures are envisioned and enacted through particular forms of bioinformation; and how are totalizing aspirations realized or disrupted? Further, how does bioinformation illuminate what we imagine life is, and how it should be fostered?

The essays in this volume that foreground critique rooted in critical race studies and feminist STS are especially illuminating here, most notably Anna Jabloner's essay "American bioinformation and U.S. race politics: the values of diverse genetic data" and Sylvia McKelvie's essay "Bioinformation and the politics of rape response." In both cases, we learn that those who amass and use databases do so in the name of particular ideas about justice that they anticipate these forms of bioinformation can advance. And yet as Jabloner and McKelvie each show, whether by design or by unexpected outcome, they can instead produce the opposite effects. Here questions of justice turn on what and who is valued, and what and who is credible.

Jabloner compares two forms of bioinformation repositories—biomedical databases and forensic ones—to understand how their custodians seek to account for and combat racist bias. Crucially, diversity and consent are what make biomedical databases valuable and credible, but neither is present in forensic databases: diversity is not present as such because these databases aim to be "race neutral" rather than representative, and consent is not present because contributors are legally compelled to submit their DNA upon arrest or conviction. Jabloner shows that these database forms produce competing universalisms in which diversity operates as critical ingredient and as a necessary foil, biocapital flourishes, and existing racialized power relations are reproduced.

McKelvie offers a feminist analysis of bioinformation that joins critiques of carceral feminism (for example Gruber 2020) with analyses of how rape kits materialize injury as they fail to produce the kinds of truth effects that advocates promise (see Kruse 2010). Of interest here is not just how bioinformation functions as evidence, but what emerges in the process.

Two issues emerge. First, there is a vast gap between perceptions of the effectiveness of forensic bioinformation in preventing rape and a reality in which most rape kits are not analyzed, most of those analyzed do not lead to convictions, and courts' preferences for DNA evidence undermine rape victims' credibility. Second, there is a deeper question that McKelvie explores about the pursuit of carceral solutions to sexual violence. Here bioinformation, which McKelvie theorizes through the idea of a body-information relation, entrenches rather than transforms existing power dynamics that render some bodies rapeable and others not. McKelvie's body-information relation foregrounds that bioinformation is always already a series of relations through which further relations proliferate.

Relationality

It is perhaps a given in anthropology that all meaning is relational. But relationality is especially compelling when considering bioinformation because it attunes us to questions of motivation, goals of standardization, and the challenges of scale. In the essays in this volume, relations abound, both in the sense of the assemblages forged through bio extraction, transfer, and organization and in the sense of the collaborative intimacies required to produce bioinformation that moves. As Van Allen reminds us, "objects are made meaningful according to how they are placed within relations of significance" (Van Allen, this volume).

Cruz, Tinkler, and Fenton, discussed above, give us a lot to think about here. Their data scavenging methodology that pulls together in new ways marginalia not meant to be part of the methodology of the long-running National Survey of Health and Development that they examine shows how new questions become possible through shifting relations between method and data. They also underscore that in many instances it is cultivated relations, here between study staff and research participants over more than 70 years, that make the production of bioinformation possible.

That question of what makes the production of bioinformation possible is not only one of the relational efforts of researchers, but also one of the motivation of participants. Here power again comes into focus, but not necessarily in ways we might expect. Kiheung Kim and Jongmi Kim's analysis of the motivation of South Koreans who voluntarily subject themselves to state surveillance as a way to track the spread of Covid-19 speaks back to others' analyses that emphasize stereotyped cultural values and the specter of growing authoritarianism. Instead of obedient citizens and authoritarian governments, Kim and Kim's analysis brings into focus a public forged through recent democracy movements and other disease outbreaks. Voluntarily providing authorities access to their biodata is not a giving up of rights, but rather an assertion of two rights: the right to government transparency, and the right to protect themselves from infectious disease that could only come with voluntary bioinformation flows. Kim and Kim's emphasis on public relationships to bioinformation is especially important given tensions between impulses within bioinformation projects towards, on the one hand, openness and access, and on the other hand towards secrecy and specialization. Jabloner's analysis of the opposed value of more and less diversity in different database forms is helpful towards conceptualizing this tension, as is the push that we see most clearly in Murphy and Maguire's and Chadha's essays to move through the space of critique to arrive instead at a place of collaboration.

Towards a future anthropology of bioinformation

I began this essay by asking why bioinformation is such a fruitful object of research in this moment, and why we need the kinds of critical approaches presented here. Ultimately the point, and the reason that this volume is so necessary and instructive, is that the more totalizing bioinformation projects aspire to be, the more necessarily unfinished they will become. The gaps between aspiration and realization, the material excesses that accompany and exceed bioinformation's representational possibilities, and the insistent relationalities that resist distillation are the sites for a present and future anthropology of bioinformation.

I'll end with the questions that these essays sparked for me, but that cannot yet be answered. What kinds of translations are attempted or stumbled into as bioinformation emerges? How does temporality become a mobile, flexible thing, and how do we approach efforts to fix time into something predictable and controllable? How do the practices of salvage and capture that emerge in a number of the papers relate to other forms of salvage and capture that co-occur in the times and places under discussion? What forms of life are being imagined across Global North and Global South, in relation to empire in decline, and in the context of the nation-state? Where and what is value as life becomes data?

References

Chadha, this volume.

Cruz, Tinkler and Fenton, this volume.

Gonzalez-Polledo and Posocco, this volume.

Gruber, Aya. 2020. *The Feminist War on Crime: The Unexpected Role of Women's Liberation in Mass Incarceration*. Berkeley, CA: University of California Press.

Jabloner, this volume.

Kim and Kim, this volume.

Kruse, Corinna. 2016. *The Social Life of Forensic Evidence*. Berkeley, CA: University of California Press.

Little, this volume.

McKelvie, this volume.

Murphy and Maguire, this volume.

Nadim, this volume.

Tsing, Anna. 2005. *Friction: An Ethnography of Global Connection*. Princeton, NJ: Princeton University Press.

Van Allen, this volume.

Zhan, Mei. 2009. *Other Worldly: Making Chinese Medicine Through Transnational Frames*. Durham, NC: Duke University Press.

Index

For Product Safety Concerns and Information please contact our EU
representative GPSR@taylorandfrancis.com
Taylor & Francis Verlag GmbH, Kaufingerstraße 24, 80331 München, Germany

www.ingramcontent.com/pod-product-compliance
Lightning Source LLC
Chambersburg PA
CBHW060301220326
41598CB00027B/4194

9 7 8 1 0 3 2 1 4 0 8 2 7